WITHDRAWN
from Toronto Public Library

SPACE EXPLORATION
ALL THAT MATTERS

D0094103

SPACE EXPLORATION

ALL THAT MATTERS

David Ashford

First published in Great Britain in 2013 by Hodder & Stoughton. An Hachette UK company.

First published in US in 2013 by The McGraw-Hill Companies, Inc.

This edition published 2013

Copyright © David Ashford

The right of David Ashford to be identified as the Author of the Work has been asserted by him in accordance with the Copyright, Designs and Patents Act 1988.

Database right Hodder & Stoughton (makers)

All rights reserved. No part of this publication may be reproduced, stored in a retrieval system or transmitted in any form or by any means, electronic, mechanical, photocopying, recording or otherwise, without the prior written permission of the publisher, or as expressly permitted by law, or under terms agreed with the appropriate reprographic rights organization. Enquiries concerning reproduction outside the scope of the above should be sent to the Rights Department, Hodder & Stoughton, at the address below.

You must not circulate this book in any other binding or cover and you must impose this same condition on any acquirer.

British Library Cataloguing in Publication Data: a catalogue record for this title is available from the British Library.

Library of Congress Catalog Card Number: on file.

10 9 8 7 6 5 4 3 2 1

The publisher has used its best endeavours to ensure that any website addresses referred to in this book are correct and active at the time of going to press. However, the publisher and the author have no responsibility for the websites and can make no guarantee that a site will remain live or that the content will remain relevant, decent or appropriate.

The publisher has made every effort to mark as such all words which it believes to be trademarks. The publisher should also like to make it clear that the presence of a word in the book, whether marked or unmarked, in no way affects its legal status as a trademark.

Every reasonable effort has been made by the publisher to trace the copyright holders of material in this book. Any errors or omissions should be notified in writing to the publisher, who will endeavour to rectify the situation for any reprints and future editions.

Typeset by Cenveo® Publisher Services.

Printed and bound in Great Britain by CPI Group (UK) Ltd., Croydon, CR0 4YY.

Hodder & Stoughton policy is to use papers that are natural, renewable and recyclable products and made from wood grown in sustainable forests. The logging and manufacturing processes are expected to conform to the environmental regulations of the country of origin.

Hodder & Stoughton Ltd

338 Euston Road

London NW1 3BH

www.hodder.co.uk

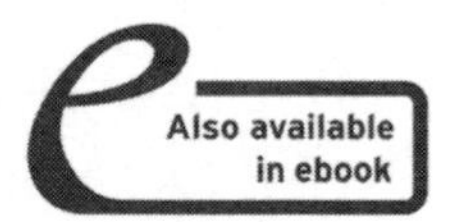

About the author

David Ashford is the Managing Director of Bristol Spaceplanes Limited, an innovative small company developing the *Ascender* spaceplane. He graduated from Imperial College in aeronautical engineering and spent one year at Princeton doing post-graduate research on rocket motors. His first job, starting in 1961, was with the Hawker Siddeley Aviation hypersonics design team, working on spaceplanes, among other projects. He has since worked as an aerodynamicist, project engineer, and project manager at Douglas Aircraft and at what is now BAE Systems on various aerospace projects, including the DC-8, DC-10, Concorde, the Skylark sounding rocket, and various naval missile and electronic warfare systems. He co-authored with Professor Patrick Collins the first serious book on space tourism, *Your Spaceflight Manual – How You Could Be a Tourist in Space within Twenty Years* (Headline, 1990), and wrote a follow-up book *Spaceflight Revolution* (Imperial College Press, 2002). He has had published about 20 papers on space transportation in the professional press. He is a Fellow of the British Interplanetary Society, a Fellow of the Royal Aeronautical Society, and is a Rolt Fellow at the Centre for the History of Technology at the University of Bath.

Declaration of interest

The author would like declare an interest in that his own company has plans to develop small spaceplanes. In a few instances, the text describes some of these projects to illustrate the points being made. This is only where they are the only suitable exemplars, and it is not suggested here that they are necessarily the best projects. The author has tried to prevent this interest from biasing the arguments used. If any reader feels that the result is unfair, by commission or omission, they are most welcome to comment. An open debate on the way ahead for space would benefit us all.

Acknowledgements

I would like to thank Dr Ian Crawford, Reader in Planetary Science and Astrobiology, Department of Earth and Planetary Sciences at Birkbeck College, University of London, for commenting on the parts of the text dealing with space science; Joanne Wheeler, Partner at CMS Cameron McKenna LLP, for commenting on the parts dealing with space law; and Professors R. W. Ashford, R. A. Buchanan, and P. Q. Collins on more general aspects. Any misinterpretation of their comments is entirely my responsibility. I would also like to thank the various copyright holders for permission to use their images.

Photographic credits

©ESA-CNES-ARIANESPACE-Optique (Fig. 2.2a); NASA (Fig. 2.2b); Bettmann/Corbis Images (Fig. 3.1); NASA (Fig. 3.2a); Courtesy of Scaled Composites, LLC (Fig. 3.2b); GKN Aerospace (Fig. 3.4); MarsScientific.com and Clay Center Observatory (Fig. 3.5a); Rex Features (Fig. 3.5b); Bombardier Aerospace (Fig. 4.1); Steve Mann/Shutterstock.com (Fig. 4.2); Morphart Creation/Shutterstock.com (Fig. 5.1a); Brad Whitsitt/Shutterstock.com (Fig. 5.1b); Corbis Images (Fig. 9.1); de Havilland (Fig. 11.1); Justin Lewis (Fig. 12.3)

Contents

1 Introduction 1

2 Why space? 10

3 A brief history of launch vehicles 18

4 Getting to space 40

5 Revolution 55

6 The new space age 60

7 Your space tourism experience 75

8 Who owns space? 83

9 Space and the environment 87

10 Space colonization 92

11 Safety 103

12 Roadmap to the new space age 111

13 Finally 132

100 Ideas 135

Glossary 146

Index 150

Introduction

If you are reading this, you probably don't need persuading that space exploration is one of the most interesting and exciting prospects facing the human race. Very large space-based instruments and long-distance robotic probes hold the keys to answering the question of whether we are the only intelligent life in the universe, which in turn is a vital part of the quest to understand our place in the overall scheme of things. Investigation of other planets should enable a far better understanding of how life evolved on our home planet. Resources from space, probably starting with collecting solar power from large satellites and mining asteroids, should help to relieve the human pressure on planet Earth. Human visits to nearby planets and their moons should add insights that we cannot gain from robots.

You are probably also feeling frustrated and puzzled. What happened to the plans for lunar and Mars explorations that were promised in the heady pioneering days of the 1960s? Stanley Kubrick's film, *2001: A Space Odyssey*, inspired by the work of Arthur C. Clarke, became the highest grossing picture from 1968 in the USA. It includes a human visit to Jupiter and, such was the rate of progress at the time, was widely thought to be a plausible indication of what might actually happen by 2001. Inspired by the public reaction to this film, Pan Am and Thomas Cook began registers of names for those interested in lunar tourism. In its heyday, the list numbered nearly 93,000. Yet nothing remotely comparable has happened or is even planned. Why, more than 40 years after a man last stepped foot on the moon in 1972, is there no coherent scheme for setting up a lunar base? How can it be that, more than 50 years after the first man visited space in 1961, it still costs tens of millions of dollars to send someone to orbit? How do the new commercial space activities, such as Richard Branson's Virgin Galactic, fit in with NASA and other space agency plans?

This book attempts to answer these questions and to derive a roadmap for speeding up space exploration at affordable cost. A major conclusion is that the biggest obstacle is the entrenched habits of thinking of NASA and the other large space agencies. In this situation, an open debate on the way ahead for space would benefit us all, and the more people who understand the basics of space exploration the better. This book aims to help with this process. The rest of this Introduction surveys the main issues and serves as a synopsis for the remaining chapters.

Until quite recently, there was optimism that we would soon have a lunar base that would be followed by human visits to Mars. This optimism was carried forward on the momentum generated in the 1960s by the race to the Moon between the Soviet Union and the USA. One of the greatest technical feats in human history (if not the greatest) was NASA's Apollo project that put 12 men on the moon between 1969 and 1972 and got them all back safely. However, this achievement was motivated more by the politics of the Cold War than by any altruistic desire to explore space. With the Cold War over, budgets are no longer adequate for a serious space exploration programme along the old lines. The remains of Apollo on Earth are museum pieces. The remains on the Moon are among the first examples of extra-terrestrial industrial archaeology. (Although you will have to wait a few years for the guided tour!) There are no funded replacements in sight.

During the Cold War, achievement was far more important than cost. It was therefore considered acceptable to carry on using launchers that could fly once only. This followed on from the use of converted ballistic missiles for pioneering spaceflights. This habit has continued ever since, and is the primary cause of the high cost and risk of space transportation. Motoring would be a very niche activity if cars were scrapped after every journey.

Clearly, high cost is the major impediment to an expanded space exploration programme. But none of the major space agencies has firm plans to reduce cost significantly.

However, there is a new way ahead, based on a commercial rather than a big government approach to space exploration. A recent example is the setting up of Planetary Resources, a company based in Bellevue, Washington, USA, to use robotic spacecraft to prospect asteroids for precious materials such as platinum. This may seem like science fiction until you look at the names of the founders and advisors. Larry Page is the founder of Google. James Cameron directed the *Titanic* (1997) blockbuster film. Eric Anderson started the Space Adventures company, which has arranged for seven private citizens (as of November 2012) to visit the International Space Station (ISS) on Russian rockets.

Peter Diamandis founded the X-Prize, which led to the first privately funded vehicle reaching space height in 2004 (the SpaceShipOne (SS1) built by Burt Rutan's Scaled Composites company). Prior to that, Diamandis had been the prime mover of setting up the International Space University.

Another 'new space' venture is Richard Branson's Virgin Galactic, which plans to carry the first passengers on brief space experience flights in 2014 or 2015. The chosen vehicle is SpaceShipTwo (SS2), which is an enlarged and improved version of SS1.

Several other private companies are planning vehicles to carry people to space, and one of them, SpaceX, has developed rockets at much reduced costs. They have started to demonstrate the technology for a reusable rocket. They made history in May 2012, when a Falcon 9 became the first privately funded launcher to send a module – a Dragon capsule also built by SpaceX – to the ISS and back. The expression 'the new space movement' is being used to describe these private sector activities.

This is not the place to comment on the prospects for these specific ventures, except to say that someone, some day, will make a commercial success of mining asteroids, carrying passengers to space, and building reusable launchers. The main point for now is that these 'new space' entrepreneurs all have a strong incentive to reduce greatly the cost of access to space – far more so than the established government space agencies, which survive well enough on the status quo. If these low costs can be achieved, they will in turn enable a rapid acceleration of space exploration, including a lunar base and visits to Mars.

However, the various new space companies are all pursuing different technical solutions, and the big government agencies such as NASA are not taking these ventures into account when planning future space exploration. So, at present, there is no consensus on a roadmap for opening up spaceflight to business, individuals, and expanded exploration.

However, back in the 1960s, there was indeed a consensus on how to build reusable rockets that would reduce greatly the cost of spaceflight. My first job was working on one of these projects. At the time, there was agreement that fully reusable launchers like aeroplanes (spaceplanes) were the obvious next step and that they were just about feasible with the technology of the time. Spaceplanes are like aeroplanes in engineering essentials. If they had been built, they would have introduced an aviation approach (technology and mindset) to replace the missile culture then prevailing. However, spaceplanes have still to enter service, and the 1960s ideas have been largely forgotten or overlooked.

This book considers present government plans, private sector space plans, and the 1960s consensus to see if a roadmap can be found that could provide greatly reduced cost of access to space reasonably soon, leading to a new space age. The conclusion is that there is such a roadmap and that costs can be slashed soon using more or less existing technology and on a modest budget.

Chapter 2 presents a snapshot of present space activities. Despite their high costs, satellites have transformed astronomy, Earth science (including meteorology and environmental science), navigation, and communications. However, high cost remains the major barrier to expanded space exploration and new commercial activities.

Chapter 3 presents a brief history of launch vehicles to explain how our present situation arose. Spaceplanes were widely considered feasible in the 1960s but were not built then because of Cold War pressures. They were seen as the obvious way to reduce launch costs. The decision in the late 1970s to build the Space Shuttle as not fully reusable has held up low-cost access to space by 30 years and counting. The habit of throwing away a launcher each time it was used then became so entrenched that the large space agencies like NASA have never given serious consideration to spaceplanes that do not require advanced technology. Chapter 3 then presents a snapshot of present plans by governments and the private sector's new space movement. NASA is still planning large new expendable launchers. The private sector is developing some spaceplanes and will almost certainly achieve the desired cost breakthrough eventually. Nevertheless, as explained later in Chapter 12, this could be achieved far sooner and at less cost by resurrecting 1960s designs. Chapter 3 also tries to explain the reluctance of space agencies today to take spaceplanes seriously.

Chapter 4 describes the design problem that most differentiates launchers from other vehicles, and how the cost of sending people to space can be reduced by roughly 1000 times by introducing spaceplanes based on 1960s projects. The result would be nothing less than a revolution in spaceflight.

Chapter 5 compares the potential revolution in spaceflight with previous revolutions in land and air transport, brought about by the invention of the steam locomotive and aeroplane respectively.

Chapter 6 presents a brief description of the 'new space age', following the revolution. There will be a new golden age of space science and exploration. Visits to space hotels will be affordable by middle-income people prepared to save for the trip of a lifetime. Space tourism is likely to become the first new market to generate the funding and enthusiasm needed to develop spaceplanes to their full potential.

Chapter 7 describes what a few days in a space hotel would be like. If you react like most astronauts to date, you will find it a transforming experience.

The next four chapters describe various other aspects of the spaceflight revolution. Chapter 8 considers some of the legal and regulatory matters. Chapter 9 describes the effects on the environment of low-cost access to space, which are mainly beneficial, and Chapter 10 speculates on some long-term prospects for the human race in the universe. Chapter 11 considers safety, which will become more important as larger numbers of people visit space and which will have a large influence on spaceplane design.

Chapter 12 puts it all together and presents a roadmap for developing the pivotal project for the new space age – the first orbital spaceplane – rapidly and at low cost and risk. The proposed roadmap involves rapidly developing 1960s-style spaceplanes and thereby in effect catching up with what might have been. The 1000 times cost reduction could be approached within about 15 years, if space tourism takes off rapidly enough to generate the required funding. Chapter 12 goes on to describe how you as an individual could trigger the spaceflight revolution by drawing attention to the best way ahead in a way that is likely to lead to further action. It is possible for an individual to make such a big difference because the basics were sorted out a few decades ago but have never been put into effect.

The Glossary defines some of the technical terms and acronyms used.

The emphasis throughout the book is on spaceplanes leading to revolution and a new space age. The importance of these developments is such that they are, in effect, **all that matters** when considering the future of space exploration at present.

Why space?

This chapter summarizes the advantages of space as a location for science and commerce and provides a snapshot of current space activities.

Astronomy

Astronomy from the surface of the Earth is limited by the atmospheric absorption of much of the radiation from galaxies, stars and planets. Figure 2.1 shows a rough plot of which parts of the radiation spectrum are most affected.

Gamma rays, X-rays, ultraviolet light, most of the infrared spectrum, and long-wave radio waves are all blocked by the atmosphere. Only radio waves at short and medium wavelengths, and some visible light, get through. That visible light is then subject to distortion from atmospheric turbulence.

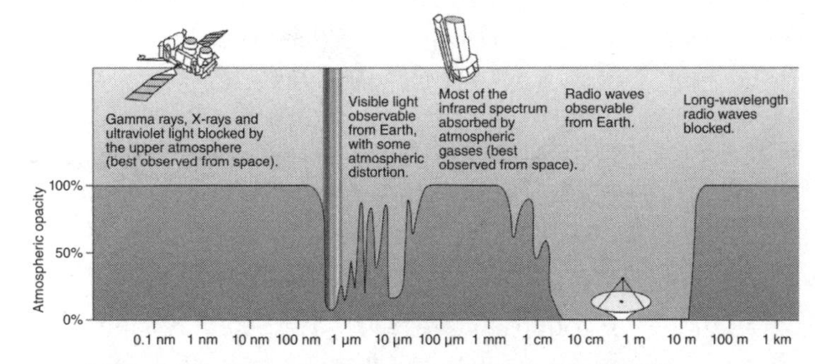

▲ **Figure 2.1 Atmospheric opacity. [Originated by NASA].**

From pre-history until the early seventeenth century, our only available instrument for observing celestial bodies was the human eye assisted by various devices for measuring angles. Then the invention of the optical telescope provided a major boost. Radio telescopes were invented in the 1930s, and the development of radar technology in the Second World War led to rapid advances in the new science of radio astronomy. However, we were still restricted to studying the radiation that could penetrate the atmosphere.

The development of satellites finally removed this restriction and among their earliest uses was carrying telescopes to measure gamma rays, X-rays, ultraviolet light, visible light clear of atmospheric distortion, and infrared and microwave radiation. It is no exaggeration to say that these measurements have enabled a transformation of our knowledge of the universe.

Satellites are playing an increasing role in detecting planets outside our own Solar System, called extrasolar planets or exoplanets. The first was discovered in 1992, and a total of 778 such planets had been identified as of 15 June 2012. The number of known exoplanets is increasing exponentially. These recent discoveries greatly increase the number of places where it is worth our while to look for life.

Spacecraft can be used to transport robotic probes or humans to other bodies in the Solar System. There is a natural sequence in the exploration of a particular heavenly body. First, remote observation using telescopes of various kinds, both terrestrial and space-based. Second, sending probes of increasing sophistication near to the body in question, often in orbit around it, to make close-up observations and transmit the results back to Earth. Third, landing a probe on the body to investigate atmospheric and surface features, again transmitting the results back to Earth. Fourth, returning samples of atmosphere, soil, or rocks to Earth. Finally, human explorers visiting as close to the body in question as is practicable, which in many cases will not be to the surface.

We have sent close-observation probes to the Moon, Venus, Mars, Jupiter, Mercury, Saturn, Uranus, Neptune and several asteroids, comets and planetary moons. We have landed probes on the Moon, Mars, Venus, Titan (Saturn's largest moon) and two asteroids. We have returned samples from the Moon, from the tail of one comet, and from one asteroid. In 2011, the Russian Federal Space Agency (Roscosmos) launched a mission aimed at returning a sample from one of Mars' moons, Phobos, but the attempt failed. According to present plans, which are not yet firm, it is likely that the first sample from Mars will be returned between 2020 and 2030. So far, we have sent humans to just the Moon. This exploration using space probes has transformed our knowledge of the Solar System.

At present, the main platform for human spaceflight is the International Space Station, which is a large satellite with accommodation for up to six scientists. However, there are no firm plans for returning to the Moon.

Other activities

Space is also widely used for military purposes such as reconnaissance, information gathering, communications, and providing links for controlling drones. By far the largest military user of space is the USA. The United States Air Force Space Command alone employs some 47,000 people.

Satellites can observe large areas of the Earth at the same time and have been used to enhance our knowledge of the atmosphere, oceans, geology, and land use. They are used for mapping and for routine global weather observations as the basis for forecasting.

Many radio wavelengths used for communication depend on the receiver being within line of sight of the transmitter and therefore cannot be used for direct long-distance wireless, television, or telephone links between points on the surface of the Earth. Satellites are routinely used as links, thereby enabling global communications coverage at these wavelengths. Hence, global news and sports coverage, satellite television, and long-distance telephone links.

The position of a satellite in orbit can be measured and predicted with precision, and this is the basis for satellite navigation. Mobile phones can now tell you where you are to an accuracy of tens of metres.

Launchers

Given the usefulness of satellites, it is perhaps surprising how few are launched. In 2011, which is typical of recent years, there were just 84 orbital launch attempts of which 78 were successful. Seven of these were manned. Twenty types of rocket were used, from eight countries. The largest in terms of payload capacity was the Space Shuttle, which made its final flight in 2011 and which was capable of launching 24 tonnes (53,000 lb) to low orbit. The smallest was the Iranian Safir, capable of launching just 27 kg (60 lb). The cost of a Space Shuttle launch averaged about $1 billion and the cost of using a small launcher can be down to a few tens of millions of dollars. This cost is far higher than that of any terrestrial transportation and is the main obstacle to expanding space exploration.

The main reason for this high cost is that all of these vehicles use large expendable components that are based on ballistic missile technology. Figure 2.2 shows a typical expendable launcher, Ariane 5, and the Space Shuttle. Ariane 5 is entirely expendable. After each launch, the various stages either crash into the sea or burn up on re-entering the atmosphere. The Orbiter stage of the Space Shuttle was like an aeroplane and was reusable. The two Solid Rocket Boosters were recovered at sea by parachute. They were then returned to the factory where some of the components were refurbished and used again. From the point of view of cost and safety, they were little better than if they had been expendable. The largest component was the External Tank, which burnt up on re-entry. As will be discussed later in Chapter 11, expendability is also the main cause of the high risk of spaceflight to date. The reasons for expendability are discussed in Chapter 4.

The development of all of these launchers was heavily subsidized by governments, usually in the interests

▲ **Figure 2.2 Ariane 5 and the Space Shuttle. The Shuttle first flew in 1981 and averaged fewer than five flights per year at a cost of some $1 billion per flight.**

of national defence or prestige. This explains the apparently inefficient use of 20 types of rocket for only 76 successful launches.

Suborbital

The above activities all require launching spacecraft into orbit. This in turn requires climbing clear of the effective atmosphere to space height and then accelerating horizontally to satellite speed, which is about 17,500 mph (7.8 km/sec) at a height of 200 km (120 miles), or round the world in 90 minutes. It is the acceleration to satellite speed that requires most of the energy – just climbing

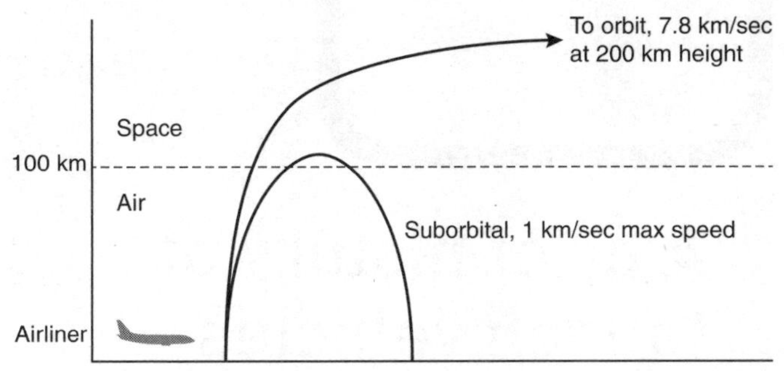

▲ **Figure 2.3 Orbital and suborbital launch trajectories.**

to space height requires far less. A so-called suborbital flight – up and down to space height with just a few minutes in space – requires a maximum speed of about 2200 mph (1 km/sec), as shown in Figure 2.3.

There is no clear boundary between the atmosphere and space, but space height is usually defined as 100km (62 miles) (a more detailed description is given in the Glossary). Expendable suborbital rockets, called sounding rockets, have been used for many years for various kinds of scientific research. They are far smaller and less expensive than orbital launchers, but only provide a few minutes in space. As will be discussed in Chapter 3, two fully reusable suborbital spaceplanes have already flown, and new ones are being developed to provide passengers with suborbital space experience flights. These will be useful stepping stones to orbital spaceplanes.

3

A brief history of launch vehicles

The pioneering phase

The first spacefaring vehicle was the German V-2 ballistic missile of the Second World War, which first reached space height in 1942 (see Figure 3.1). The V-2 was a suborbital vehicle, spending little time in space before re-entering the atmosphere and delivering its warhead. It had the now-classic configuration. The payload, in this case a high-explosive warhead, was at the front. Most of the body was filled with rocket propellant, in this instance liquid oxygen as the oxidizer and alcohol diluted with water as the fuel. The rocket engine was at the base. This was where the oxidizer and fuel reacted to produce a hot gas at high pressure, which then emerged at high speed through the nozzle to produce the thrust.

More than 3000 V-2s were launched, mostly against Antwerp and London, in the last year of the Second World War. It was a brilliant weapon technically, but it

▲ Figure 3.1 The German V-2 ballistic missile of the Second World War. This is the progenitor of all ballistic missiles and launch vehicles since then. It could fly once only.

was not very effective because of poor reliability and accuracy. The only target that it stood much chance of hitting was a large city and even there it was unlikely to damage more than a few buildings.

The nature of its task meant that the V-2 was expendable, although a piloted reusable bomber version, with wings for landing, was designed. However, the war ended before this version could be built.

The V-2 was years ahead of any competition, and formed the basis for post-war US and Soviet ballistic missile development. The aim was to overcome the limitations of the V-2 by combining more accurate guidance systems with nuclear warheads. The first indigenous US ballistic missile, the Redstone, was in effect a re-engineered and enlarged V-2. Its design team was led by Germans recruited to work in the USA, the best known being Wernher von Braun.

Because ballistic missiles can fly to space, it was natural to use them to launch the first satellites. Redstone formed the lower stage of the four-stage Juno 1, which in 1958 became the first US vehicle to launch a satellite into orbit. This was in rapid response to the first ever satellite – the Soviet Sputnik launched in 1957. In 1961, Redstone also launched the first American into space, Alan Shepard, again in response to a Soviet first when Yuri Gagarin became the first man in space earlier that year. There was actually a big difference between these two achievements – Shepard's flight was suborbital whereas Gagarin's was fully orbital. Achievement in space became a key part in the propaganda battle between the Soviet Union and the USA during the Cold War.

The race to the Moon

The next major step after sending men to orbit was the race to the Moon. The US effort was kicked off by President Kennedy's famous speech In May 1961, just six weeks after Gagarin's flight to space, when he galvanized the nation by saying:

> ***"If we are to win the battle that is going on around the world between freedom and tyranny, if we are to win the battle for men's minds ... I believe that this nation should commit itself to achieving the goal, before this decade is out, of landing a man on the Moon and returning him safely to Earth."***

The first sentence in the quotation shows how political the objectives were. Kennedy was far more interested in beating the Soviets than in exploring space for its own sake. Project Apollo was a brilliant success: 12 men stood on the Moon between 1969 and 1972, and all returned safely. The Soviet effort was an expensive failure. The West went on to win the Cold War, Apollo having played its part.

To save time in the race to the Moon, the mighty Saturn series of large new launchers was built as expendable. Von Braun had earlier proposed a large reusable vehicle, using the experience of two experimental V-2s that had flown with wings in 1945, but there was not enough time to develop these and to be reasonably sure of reaching the Moon before the Soviets.

Rocket-powered research aeroplanes

In parallel with the development of larger and more capable satellite launchers, the USA led the way in developing a series of rocket-powered high-speed research aeroplanes. The Bell X-1 became the first piloted aeroplane to exceed the speed of sound (Mach 1) in 1947; the Douglas Skyrocket reached Mach 2 in 1953; and the Bell X-3 reached Mach 3 in 1956. The last in this series was the North American Aviation Inc. X-15 (see Figure 3.2), which made its first flight in 1959 and its last in 1968. The X-15 was designed to explore hypersonic speeds and it eventually reached Mach 6.7. Perhaps more importantly, it was also designed to bring in an aviation approach to spaceflight by using rocket-powered aeroplanes instead of expendable converted ballistic missiles. Many of the X-15 team were driven by a vision of everyday economical transport to space. The X-15 could reach space height

▲ Figure 3.2 The X-15 and SpaceShipOne (SS1) – the only suborbital spaceplanes to have flown. Thirty-six years elapsed between the last spaceflight of X-15 in 1968 and the first of SS1 in 2004.

and was the first true suborbital spaceplane. It could have been used as the reusable lower stage of a satellite launcher, but a feasible proposal to do this was rejected due to concerns over safety and economics. As well as carrying out invaluable aeronautical research, the X-15 carried several space science experiments, serving as a reusable sounding rocket. Jumping ahead in our story, the Scaled Composites SpaceShipOne (SS1), also shown in Figure 3.2, was the next aeroplane to reach space height. It achieved this feat in 2004, 36 years after the X-15 last flew to space. This hiatus is an indication of the lack of priority given to reducing the cost of access to space.

1960s spaceplane studies

In parallel with the Apollo programme, and partly inspired by the X-15, most large aircraft companies in Europe and the USA worked up proposals for fully reusable launchers that could fly all the way to orbit, which would now be called orbital spaceplanes. At a Society of Automotive Engineers (SAE) space technology conference in Palo Alto, California, in 1967, for example, which I attended, no fewer than 15 proposals for reusable launch vehicles were presented, and there were several others not represented there.

Among these were eight designs prepared by European companies as part of the so-called Aerospace Transporter project. This programme was coordinated by Eurospace under the guiding mind of Professor Eugen Sänger, who had designed a rocket-powered suborbital bomber in the Second World War and who can well be described as the father of the spaceplane. These spaceplane designs were all relatively small, with a payload of 1 or 2 tonnes to orbit. Most of the US designs were considerably larger.

There was a consensus that reusable launchers were the obvious next step in space transportation. Consuming a complete vehicle for each launch could never be economical. There was also a consensus that such vehicles were feasible with the technology of the day. The suborbital X-15 was demonstrating that the technical problems of orbital spaceplanes were within sight of being solved.

With just a few dissenters, there was a further consensus on the essential design features, and most of these 1960s projects had the following in common.

- More than one stage, so that the technology of the time could be used.
- The use of recently developed hydrogen-fuelled rockets, which had significantly improved performance.
- Wings to provide lift for landing, as being safer and more practicable than rotors, vertical jets, vertical rockets, or parachutes.
- Pilots, as being safer than autopilots or remote control.

These design features will be considered in more detail in Chapter 12, which will show that they are as competitive now as then.

The biggest unknowns at the time concerning spaceplane design were the effects of wings on re-entry stability and heating, and how to protect the structure from the heat of re-entry. The Space Shuttle has since provided a full-scale demonstration of solutions to these problems, and there have been major advances in these and other technologies since the Shuttle was designed. All the technologies for an orbital spaceplane have now been demonstrated in flight.

Figure 3.3 shows the design from the 1960s that, in my opinion, has best stood the test of time – the French Dassault Aerospace Transporter – although several of the others come very close (such as the British Aircraft Corporation MUSTARD; the Douglas Astro; the Junkers Sänger 1; the Lockheed Star Clipper; the Martin Marietta Astrorocket; and the McDonnell Douglas Tip Tank project).

The French Dassault Aerospace Transporter was fully reusable. The lower stage (Carrier Aeroplane) was a large

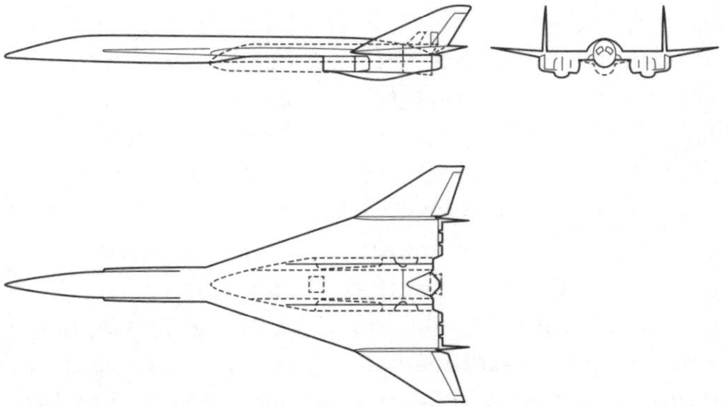

▲ Figure 3.3 The Dassault Aerospace Transporter of the mid-1960s.

supersonic aeroplane using jet engines to accelerate to Mach 4. The rocket engines on the upper stage (Orbiter) were then started, and the combination accelerated to Mach 6, at which speed the Orbiter separated and made its way to orbit while the Carrier Aeroplane flew back to base. Some of the spaceplane projects proposed in later chapters are generally similar to this design.

These 1960s spaceplane projects were designed mainly for launching satellites and sending people and supplies to space stations. However, they could have been adapted for carrying passengers on long-distance flights or for space tourism. If a spaceplane accelerates to just short of satellite speed, it can then glide halfway round the world. The flying time from Europe to Australia, for example, would be about 75 minutes. At the time, space tourism had hardly been considered seriously, and fast long-range transport received more attention. A few designers thought that this would be by far the largest market for spaceplanes and that the resulting economies of scale would reduce the cost per seat to just a few times more than the fare in a conventional airliner. This raised the prospect of a fleet of a few dozen spaceplanes each making several flights per day, leading in turn to airline-like operations. The result would be an aviation approach to space transportation to replace the missile one. These studies were the first to indicate the low-cost potential of spaceplanes when used in significant numbers, and this is a key theme discussed in later chapters. One of the first research reports on these lines was *The Rocket-*

propelled Commercial Airliner by Walter Dornberger, published in 1956 by the University of Minnesota. This was the year before the first satellite launch. Earlier, as a Major-General in the German Army, Dornberger had led the V-2 programme.

Contemporary designers generally accepted this view as an interesting long-term prospect rather than as an immediate priority. However, if a spaceplane had been built in the 1960s for launching satellites, it would no doubt have been adapted for experimental long-range transport and for pioneering space tourism. If there had turned out to be a large market for either of these uses, it is likely that regular commercial operations would have started by the late 1970s. Now, many consider that the first large market for spaceplanes is more likely to be space tourism. These two applications of spaceplanes are considered further in Chapter 6.

Space Shuttle

These three strands of development – improved expendable launchers, higher performing rocket-powered research aeroplanes, and spaceplane studies – came together for the next major project after Apollo, which was the Space Shuttle. At the time, the Space Shuttle was so promising that there was great optimism about the future of space. Gerard O'Neill even proposed building space colonies in orbit, using materials from the Moon and asteroids, and these ideas gained much publicity. This was shortly after the film *2001: A Space Odyssey* (1968), which added to the general optimism.

When Shuttle design was started in the 1970s, the early proposals were fully reusable, taking advantage of X-15 experience and the numerous 1960s spaceplane studies. They were also very large, having a 30-tonne payload to meet the requirements of the US Department of Defense. Then President Richard Nixon cut NASA's budget, and the large fully reusable design could no longer be afforded. NASA then faced a choice. They could either make the vehicle far smaller but still fully reusable, along the lines of earlier projects (especially the European Aerospace Transporters), or they could give up on complete reusability. The habit of expendability was by then strong enough for NASA to make the great mistake of choosing the latter course.

As a result, the Space Shuttle, Figure 2.2, was just as expensive to fly in terms of cost per tonne of payload as the Saturn that preceded it. This put paid to several promising schemes for exploiting low-cost space transportation, although it took several years for the limitations of the Shuttle to become widely appreciated. Pilot schemes for O'Neill's space colonies, for example, might just have been affordable with the reusable Shuttle originally planned but did not stand a chance with the largely expendable one. At the time of the Shuttle decision, NASA did not seem to appreciate the profound difference between an expendable and a reusable design. The former is fundamentally unsuitable for large routine markets; the latter can be developed to service them.

In spite of this, the Shuttle was the mainstay of the US human spaceflight programme for 30 years, from its first flight in 1981, and did much useful work. However, the decision to make it not fully reusable has delayed low-cost access to space by 30 years and counting. A great opportunity was missed to bring in an aviation culture to replace the missile one. Most of the 1960s spaceplane designs would have been far better than anything built since then, or even proposed seriously by a major company.

NASA spaceplanes

NASA has not completely ignored spaceplanes since the Shuttle entered service. In the late 1980s and 1990s, it funded studies of two very advanced projects, the X-30 and X-33. Both were designed as single-stage vehicles capable of flying to and from orbit. Single-stage launchers need far more advanced technology than two-stage ones, as will be discussed in Chapter 4. The X-30 required jet engines capable of very high speed and the X-33 required a very light structure. Both turned out to be unfeasible and more than $1 billion was spent before these projects were cancelled. It is not at all clear why NASA did not instead resurrect one of the two-stage 1960s projects, which could have been built with existing technology.

Jets and rockets

Spaceplanes will need long-life rocket engines. Present high-performance rocket engines have lives of only a few tens of flights, whereas jets can endure for thousands of flights. It is therefore relevant to consider briefly some aspects of the history of jet and rocket development. Experimental jet engines were running in the late 1930s. They entered quasi-experimental operational service in British and German fighter aeroplanes in 1944 in the closing stages of the Second World War. The first jet airliner service started in 1952 with de Havilland Comets operated by the British Overseas Airways Corporation (BOAC). So, it required eight years for jet engines to mature from early operations to airline use. At that time, jet engine development was a priority in the leading aviation countries.

In terms of life and reliability, rocket engines today are arguably comparable in maturity to jet engines in the late 1940s, at least in terms of the total hours flown. There are two reasons why they have short lives. First, rocket engines operate at higher temperatures than jets. Second, there has never been a strong commercial or military incentive to develop a long-life rocket engine. They have been used mainly for missiles, expendable launchers, and a few experimental aeroplanes, none of which needs engines with long lives. The one exception is the Messerschmitt Me 163 rocket fighter, which was a contemporary of the first jet fighters. It too entered service in the closing stages of the Second World War. It was by far the fastest aeroplane in that conflict but achieved only limited success. However, the war ended before the engine approached maturity and the Me 163 remains to this day the only rocket-powered aeroplane built in significant numbers.

Operational rocketplanes

As well as the rocket-powered research aeroplanes mentioned earlier, three post-war rocket-powered aeroplanes have flown that were designed for operational use. The first of these was the French Sud-Ouest Trident rocket fighter, which first flew in 1953. Twelve were built and it came close to being selected as the next French interceptor fighter.

The second was the British Saunders Roe SR.53 rocket fighter that first flew in 1957 (see Figure 3.4). Two prototypes were built before the project was cancelled because of changed operational requirements. Its de Havilland Spectre rocket engine used technology developed from the engine used in the Me 163.

The third rocketplane was the NF-104A, which was basically a Lockheed F-104 Starfighter jet fighter with an added rocket engine. Three aeroplanes were modified, and it served with the US Aerospace Research Pilots School between 1963 and 1971 as an astronaut trainer. It was in one of these that Chuck Yeager, who earlier had been the first pilot to exceed the speed of sound, had a near-fatal accident, as shown dramatically in the film version (1983) of Tom Wolfe's excellent book, *The Right Stuff* (1979).

Between them, these three rocketplanes showed that routine flying to near-space height was quite feasible. Their rocket engines did have a reasonable life, but this was at the expense of the performance needed for efficient flight to orbit.

▲ **Figure 3.4 Saunders Roe SR.53 rocket fighter, which first flew in 1957. It did not enter service, but could have been developed into a mature aeroplane with suborbital performance.**

Although all three rocketplanes were capable of great height, actual space capability was proposed for only one. When the SR.53 was cancelled as a fighter, Saunders Roe proposed a space research conversion, to be air-launched from a Valiant bomber and capable of suborbital flight. This proposal did generate considerable interest, but not enough to make it happen.

It is interesting to speculate on what might have happened if the SR.53 had entered operational service as a fighter. Within a few years there would have been a reliable and mature aeroplane in service readily adaptable for suborbital flight to space. A two-seater derivative could have been built, and suborbital space tourism could have started in the 1960s! This would have led naturally to one of the orbital spaceplane projects being studied at the time and we would now be well into the new space age.

In the event, such was the low priority given to developing reusable vehicles that it was not until 2004 that another aeroplane reached space height – the SS1 shown earlier in Figure 3.2. This was built by Burt Rutan's Scaled Composites company and funded by Paul Allen of Microsoft fame. It won the $10 million X-Prize for the first privately funded suborbital vehicle. This prize was inspired by Peter Diamandis and, as intended, has made a major contribution towards low-cost access to space.

Historical overview

A great opportunity was missed in the 1970s to build a fully reusable orbital spaceplane. If this had happened, space transportation to orbit would now be routine and affordable. The Space Shuttle was sold originally as a reusable vehicle but, being in the event largely expendable, could never live up to its promise. This history has created institutions and habits of thought that have repeatedly reinforced the expendable habit. Even today, the big space agencies are nearly all promoting expendable launchers. This mindset remains the biggest obstacle to building an orbital spaceplane.

Present plans

NASA and other space agencies

How well do present space plans reflect the promise of the 1960s spaceplane studies described earlier in this chapter? None of the space agencies is showing much interest in spaceplanes (except possibly for the UK Space Agency, which is funding small-scale studies), although NASA is helping to fund most of the US private sector initiatives described later in this chapter. NASA's main preoccupation regarding space transportation is finding the budgets needed to build a very large new expendable launcher, SLS (Space Launch System), which, when fully developed, would be larger than the mighty Saturn of the 1960s. A major risk facing this project is that even before it is built, spaceplanes (and perhaps larger launchers using spaceplane technology for reusability) may have made it obsolete.

There seems to be no rational reason for NASA and other space agencies to be so reluctant to consider seriously the prospects for spaceplanes. The entrenched habits of thought mentioned above are one explanation. A second explanation is that a cultural gap has grown up between the space and aviation industries. All of the European and many of the US spaceplane projects of the 1960s were carried out by aeroplane design teams, with perhaps some input from engineers with experience of launchers or re-entry vehicles, but these teams were allowed to disband. Today, not many launcher designers really understand aeroplanes, and vice versa. Much of the analysis described in this book involves applying the techniques of airliner conceptual design to launch vehicles, but this approach went out of fashion several decades ago when the big aerospace companies stopped studying spaceplanes. I suspect that most *aeroplane* company design teams today, if tasked with working out a strategy for achieving the new space age soon and at affordable cost, would arrive at conclusions similar to those in this book.

A third explanation may be that introducing spaceplanes involves a complete re-conceptualization of our approach to spaceflight, and that large monopolistic government agencies are not usually among the prime movers of radical change. Three major changes will happen at more or less the same time. First is the change from launchers like missiles to those like aeroplanes, with the accompanying change in culture. Second is the change in government role from taking the lead to supporting and regulating the private sector. Third is the change in markets to dominance by large new commercial activities, especially tourism.

Whatever the explanation, if sensible evolution is suppressed for long enough, revolution follows. Progress will be rapid indeed as soon as the mindset changes.

Private sector

The most important innovations at present are from the US private sector. The leading company in terms of operational hardware is SpaceX, founded by dot.com billionaire Elon Musk. SpaceX has developed conventional-looking expendable launchers at a fraction of the traditional cost. As mentioned in the Introduction, in May 2012 a Falcon 9 made history by becoming the first privately funded launcher to send a module – a Dragon capsule also built by SpaceX – to the International Space Station and back. SpaceX has started testing an experimental reusable version, called Grasshopper, which uses its rocket engines to land vertically on a pad.

Not far behind in terms of timescale is The Spaceship Company – a joint venture between Scaled Composites and Richard Branson's Virgin Galactic. It is developing the SpaceShipTwo (SS2) spaceplane that can carry six passengers on suborbital flights, and hopes to start carrying passengers, at fares of $200,000, from 2014 or 2015. SS2 itself is launched from a large carrier aeroplane.

XCOR is developing the Lynx two-seat suborbital spaceplane. This has just one stage. In overall concept it is not unlike the SR.53 mentioned earlier in this chapter, but it has an additional seat and does not have a jet engine. (The SR.53 had a jet engine as well as a rocket.)

Bigelow Aerospace, founded by the property billionaire Robert Bigelow, is developing large commercial space stations using inflatable structures. The company has already sent early development versions to orbit.

Planetary Resources is looking into the feasibility of mining asteroids.

The Sierra Nevada Corporation is developing the Dream Chaser, which is a spacecraft of lifting body configuration to be carried by by an expendable Atlas launcher and is designed to carry tourists to orbit.

Stratolaunch Systems, founded by the aforementioned Paul Allen, is building an air-launch system with a giant aircraft as the first-stage launch platform.

This list of 'new space' activities by US startup companies is by no means exclusive. The sheer number of people in the USA combining vision with wealth, and the supporting culture and infrastructure enabling them to just get on with it, is unmatched anywhere else on this planet, and this goes a long way to explain the US lead in so many pivotal new developments.

Among established major players, Boeing is building the CST-100 capsule designed to be launched by an expendable Atlas and to carry cargo and eventually astronauts to the ISS. Recovery will be by parachute. Orbital Sciences is building a transportation system using its Antares expendable launcher, Cygnus manoeuvring spacecraft, and a Pressurized Cargo Module developed by Orbital's industrial partner Thales Alenia Space.

Despite NASA's preoccupation with SLS, they have supported most of the above projects.

Outside the USA, two projects have gained attention. The Skylon reusable launcher is being researched by the UK start-up company Reaction Engines. Skylon is a single-stage-to-orbit vehicle using an advanced new engine that combines features of jets and rockets. It is not piloted. It requires advanced technology, but is a more feasible proposition than the X-30 and X-33 mentioned earlier.

The European EADS Rocketplane is designed to carry four passengers on suborbital flights. It has jet and rocket engines and is piloted. It is not yet funded.

First orbital spaceplane

What then are the prospects of present plans leading to the new space age? Considering the above 12 projects, it is perhaps surprising that they are all very different and that not one is a piloted orbital spaceplane along the lines of the 1960s proposals. Three of these projects – the Virgin Galactic SpaceShipTwo, the XCOR Lynx, and the EADS Rocketplane – are indeed fully reusable

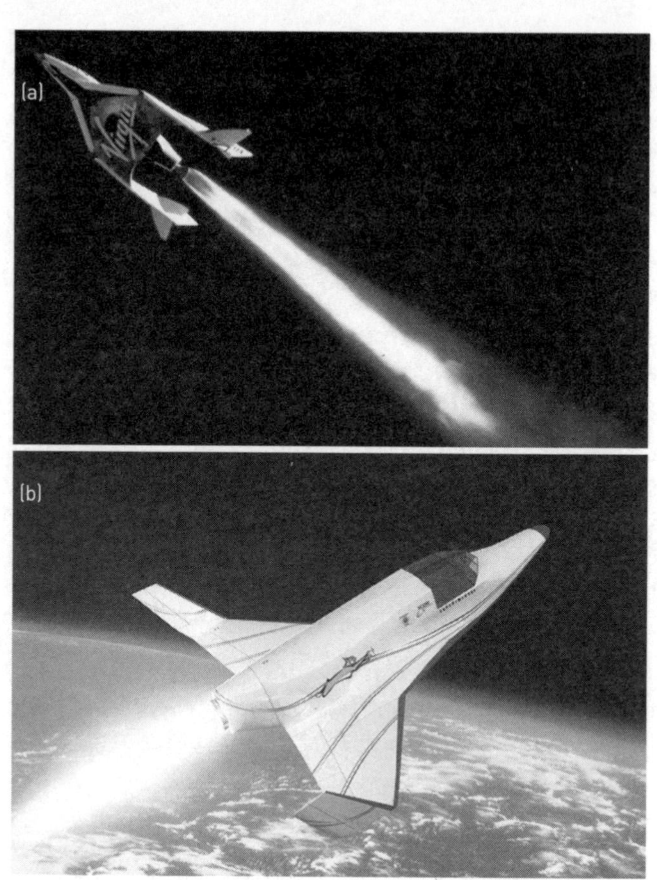

▲ **Figure 3.5 Harbingers of revolution: the Virgin Galactic SpaceShipTwo (SS2), in flight (a) and an artist's impression of the XCOR Lynx (b) suborbital spaceplanes.**

piloted spaceplanes, but these are only suborbital. Nonetheless, they would serve as a useful interim step towards the first fully orbital spaceplane, which, as will be discussed in Chapter 5, is the key to the new space age. Two of them – the Virgin Galactic SS2 and the XCOR Lynx – are funded, and these are shown in Figure 3.5.

It seems likely that one or both of these projects, and/or a competitor not yet funded, will eventually lead to commercially successful suborbital passenger flights. With economies of scale, maturing technology, and competition, the fare will come down from the present $100,000 to $200,000 to a few thousand dollars. There will be at least several dozen suborbital spaceplanes in service, each making several flights per day to space. The advantages of aeroplanes over missiles will then be clear for all to see, and the case for a fully orbital spaceplane will become unanswerable. One or more of these companies will probably then propose such a vehicle and a government space agency is likely to provide support. In this way, it seems likely that the development of the first orbital spaceplane will start within a decade or two, and the new space age can then get under way. This process could be greatly speeded up by planning for it now. The next chapter discusses the engineering basics, and Chapter 12 presents a roadmap.

4

Getting to space

This chapter explores the design challenge that most differentiates launchers from other vehicles and how this has affected their development. It then describes how the cost of access to space can be greatly reduced.

Propellant weight

The design of launch vehicles is dominated by propellant weight to a greater extent than any other form of transportation. Propellant is the fuel and oxidizer carried aboard that burn together in the rocket engine. The basic reason for needing a high propellant weight is that there is no air in space. Launchers therefore have to carry their own supply of oxidizer, as well as fuel. The only practicable type of spacefaring engine available that uses on-board oxidizer and fuel is the chemical rocket. Familiar engines that are used to power vehicles – steam, petrol, diesel and jet – use oxygen from the atmosphere as the oxidizer and the vehicles themselves have to carry just the fuel. Jet engines can therefore be used only for the early part of a spaceplane ascent to orbit that is in the atmosphere.

Jet airliner engines on take-off move typically 60 pounds (27 kg) of air for every pound (0.45 kg) of fuel consumed. Only a small portion of this is burned with the fuel but, clearly, if an airliner had to rely on liquid oxygen carried on board instead of air from the atmosphere, it would have to carry far more propellant. Typically, rocket engines use between ten and twenty times more propellant than jets, for a given thrust and time. Thus, a rocket fighter might have a duration under full rocket power of five minutes, compared with the hour or so of a jet fighter.

As examples of maximum loads that can be carried by various types of transport, a soldier can carry a backpack weighing up to about 40% of his or her loaded weight (taking as typical a maximum load of 120 lb (54 kg) carried by a 170-lb (77-kg) soldier). A large road tanker truck can carry about 65% of its loaded weight as oil, the remaining 35% being the empty weight of the truck itself. A large ocean-going oil tanker can carry about 80% of its total weight as oil. A humble aluminium beer can contains about 95% of its total weight as beer, the remaining 5% being the can.

The significance of high propellant weight on launcher design can perhaps best be explained by comparison with long-range aeroplanes. Clearly, the further an aeroplane flies, the more fuel it has to carry. Most of the work done by the engines is used to overcome air resistance. A modern long-range airliner has a maximum fuel load of typically 50% of its take-off weight and can fly just about halfway round the world. The remaining 50% is the weight of the wings, tail, fuselage, landing gear, engines, equipment, furnishings, crew, passengers and freight. A tanker aeroplane, designed for refuelling other aircraft in flight, can carry a fuel load of somewhat more than 60% of its take-off weight. Clearly, there is a limit to how much fuel an aeroplane can be designed to carry. The present record is the 82% of take-off weight held by the Virgin Atlantic Global Flyer. This was a one-off design carrying only a pilot that flew round the world non-stop in 2005 to gain the record for distance flown. This 82% is close to the practical limit for a reusable flying machine using existing technology, be it an aeroplane or a spaceplane.

Single-stage launchers

Turning back to launchers, the faster they have to go, the more propellant they have to carry. Most of the work done by the engines is used to accelerate the vehicle. Liquid hydrogen has the highest energy per unit weight of any practicable rocket fuel. A single-stage launcher accelerating to satellite speed using hydrogen-fuelled rocket engines would need to carry about 87% of its take-off weight as propellant, and this leaves only 13% for structure, engines, equipment and payload. This is well above the 82% propellant weight limit for a practical reusable vehicle. Thus a single-stage *reusable* rocket-powered launcher is beyond the present state of the art. Expendable launchers do not require the components needed for recovery, such as wings and landing gear, and can have propellant weight fractions of up to just over 90%. So single-stage *expendable* launchers are right on the margins of feasibility, although none has yet flown.

Two-stage launchers

The solution is to use more than one stage. Thus, the upper stages of Ariane 5 and the Space Shuttle, shown earlier in Figure 2.2, are helped on their way by large solid rocket boosters that are discarded when their propellant is used up. The speed increase required from the upper stages is thereby reduced and they can therefore have a lower propellant weight fraction.

A two-stage spaceplane using hydrogen-fuelled rockets can have a propellant weight fraction of 64% on each stage, which is well within the scope of existing technology. This is because a vehicle with a 64% propellant weight fraction can achieve one half the speed increase needed to reach orbit. If that vehicle then released a smaller one that also had a 64% propellant weight fraction, the latter vehicle could reach satellite speed. (In practice, it is usually better for the lower stage to have a lower propellant weight fraction and the upper stage a higher one.)

The penalty of using two stages is a low payload as a percentage of the take-off weight. For example, consider a two-stage launcher, each stage carrying 10% of its own loaded weight as payload. The upper stage forms the payload of the lower stage and has a weight of 10% of the total take-off weight. The payload of the upper stage, which is the useful payload, is 10% of that, or just 1% of the total take-off weight.

Using earlier rocket engines, an even higher propellant weight fraction would have been required. With V-2 engines, a single-stage launcher would have required no less than 98% of the launch weight to be propellant, which is clearly not feasible. The V-2 itself was only suborbital and had a 73% propellant weight fraction. The X-15 and SpaceShipOne (SS1) suborbital spaceplanes, with a broadly comparable performance, had more efficient engines and therefore needed less propellant. Suborbital spaceplanes are now well within the state of the art.

Considerations of propellant weight explain why early satellite launchers were expendable. There was insufficient weight in hand for the wings, landing gear, and other recovery equipment that could otherwise readily have been added to the expendable converted ballistic missiles then in use (as was demonstrated by two V-2s fitted experimentally with wings and flown in late 1944 and early 1945). By the early 1960s, hydrogen-fuelled rocket engines had made two-stage reusable launchers a feasible proposition, and this was the basis for the numerous spaceplane studies mentioned in Chapter 3.

A similar problem was faced by aeroplane designers in the 1930s when trying to build an airliner that could fly non-stop across the Atlantic. The required fuel weight fraction was just too high for a practical aeroplane using the technology of the time. One solution was to use two stages. Figure 4.1 shows an interesting aircraft from the late 1930s, the Short-Mayo Composite, which used this technique. A long-range seaplane was air-launched from a larger flying boat (modified from an airliner), enabling it to take off with a higher load of fuel and fly further. In 1938, it established a record distance flight for a seaplane of 6045 miles (9726 km) from Dundee in Scotland to Alexander Bay in South Africa. Although it was designed for commercial trans-Atlantic operations and made several successful trans-Atlantic flights, the advent of the Second World War resulted in only one prototype being built. It had a useful payload of 1000 lb (454 kg), which, in line with the above comments on the penalties of using two stages, is very low for the total take-off weight of just less than 50,000 lb (23,000 kg). This example illustrates that two-stage aeroplanes, and hence spaceplanes, are a practicable proposition. With the availability of somewhat more advanced technology, Pan Am started the first regular trans-Atlantic commercial operations in 1939 using Boeing *Clipper* flying boats.

Long-term solutions

Single-stage spaceplanes are clearly preferable in the long term. The most likely solution is to use jet engines for the early part of the ascent to lower the required propellant weight. However, engines using existing technology are limited to about Mach 4. From that speed to satellite speed (about Mach 25), the jets would not work and would have to be carried as deadweight. This penalty outweighs the benefit of the lower propellant weight fraction. The answer is to develop jet engines capable of higher speed, but these need very advanced technology and would take a decade or more to develop.

▲ **Figure 4.1 The Short-Mayo Composite Aeroplane of 1938.**

The development of successful two-stage spaceplanes and the resulting large new markets would provide a commercial incentive to develop such an engine. Existing jet engines *can* be used on the carrier aeroplane of a two-stage launcher because they are deadweight from their maximum speed up to only the separation speed, and do not have to be carried all the way to orbit.

The cost today

High cost is clearly the main impediment to expanding space exploration. To transport a scientist from Europe or the USA to and from Antarctica costs perhaps $30,000. At present, NASA is paying the Russian Federal Space Agency (Roscosmos) about two thousand times more (about $60 million) to carry astronauts to the International Space Station (ISS). The seven private citizens who have so far visited the ISS have paid up to $30 million each, which is perhaps more representative of the true commercial cost. With costs that high, it is not surprising that only just over 500 people have been to space since the first, Yuri Gagarin, in 1961. This is an average of just ten people per year.

The use of expendable launch vehicles is the main cause of this high cost as a new vehicle is needed for each flight. More than that, expendable launchers are inherently unsafe for carrying people and this imposes large costs in trying to get astronauts back safely. (The reasons for this are explained in Chapter 11.)

Conditions for low cost

In order to reduce cost, space transportation has to become much more like air transportation. This in turn requires fully reusable launchers. No matter how efficiently expendable launchers are made and operated, they cannot offer greatly reduced costs.

The other main condition for low cost is that the market should become large enough to require a fleet of a dozen or more spaceplanes, each making several flights per day. This is needed to provide economies of scale and to enable an aviation culture and mindset to take over. This in turn requires that the new markets made possible by spaceplanes should generate a demand for launches far larger than that of today. This condition is discussed in Chapter 6, which suggests that space tourism is likely to be the first such market.

Potential for low cost

Assuming for now that new markets do indeed emerge to provide economies of scale, how low might the cost per seat in a spaceplane to orbit become? A good starting point is to compare launchers with airliners, a typical example of which is shown in Figure 4.2. An economy seat booked in advance for a long-distance flight in an airliner costs typically about $1000. A trip to orbit costs about 30,000 times more.

Airliners are of course fully reusable. For a long-distance flight, the marginal cost of crew, fuel, maintenance, and airport charges is approximately 1000 times less than its production cost (typically $200,000 compared with $200 million for a large airliner). Much the same applies to your car – roughly speaking, the marginal cost of a few hours driving is 1000 times less than the cost of the car when new. So, if your car could be used for only a few hours before being scrapped, motoring would cost about 1000 times more than it does. The marginal cost of a launch with an expendable vehicle is of course that of a new launcher plus the cost of preparing it.

Largely because of the low cost, air travel has become a mass occupation. Some 25,000 airliners make a total of about 30 million flights per year, carrying about 3 billion passengers. The airline industry is mature and sophisticated. Manufacturers strive to improve fuel efficiency or reliability by just a few percentage points,

▲ **Figure 4.2 A typical modern airliner, the Boeing 747.**

or to shave just a minute or two from the turn-around time between flights, or to add capability to the in-flight entertainment system. What a contrast with space transportation! There are fewer than 100 launches per year to space. Each launch requires months of preparation, and satellites usually take several years to design and build.

Design changes

What design changes are needed to an airliner so that it can fly to and from orbit? These changes are in two categories. First are the changes that are unavoidable because of the need to fly to very high speed, and second are optional changes made for other reasons. This section considers the effect of the unavoidable changes on cost. The optional changes do not greatly affect cost and are considered later in Chapter 12.

Compared with airliners, we saw earlier that if existing technology is to be used, considerations of propellant weight dictate that spaceplanes must have two stages and use hydrogen fuel. Having two stages clearly increases complexity and hence cost, and hydrogen is more expensive than kerosene, the latter being the fuel most widely used for airliners.

Spaceplanes will also need to use rocket engines for at least the later part of the ascent to space where there is not enough air for jet engines. As already mentioned, present-day high-performance rocket engines have useful lives of just a few tens of flights whereas airliner jet engines endure for thousands of flights. Thus, early spaceplanes will have a high rocket engine maintenance cost. However, this maintenance cost should approach airliner levels after a long production run and a programme of continuous product improvement leading to long-life engines.

Spaceplanes will also need additional systems for operating in space and for re-entry. These include reaction controls for use outside the atmosphere, where conventional aerodynamic controls do not work, and a layer of surface insulation to protect the structure from the heat of re-entry (or advanced materials that can withstand the high temperature). These systems will increase cost, but not by very much.

The shape of the spaceplane will have to be changed to provide stability and controllability over a wide speed range. The Orbiter stage of the Space Shuttle, shown in Figure 2.2, is a good example of how this can be done. Changing shape by itself need hardly affect cost.

One factor that will reduce cost is that the Carrier Aeroplane stage has a flight time of about one hour and can therefore make more flights per day than a long-range airliner. The number of flights per day that the Orbiter stage can make will depend on how long it has to wait for a launch slot suitable for making rendezvous with the space station, and this will vary according to the type of orbit used by the latter and the location of the spaceport. In ideal circumstances, the number of flights could be several per day.

To show an example of a spaceplane with the above features and designed to be as much like an airliner as

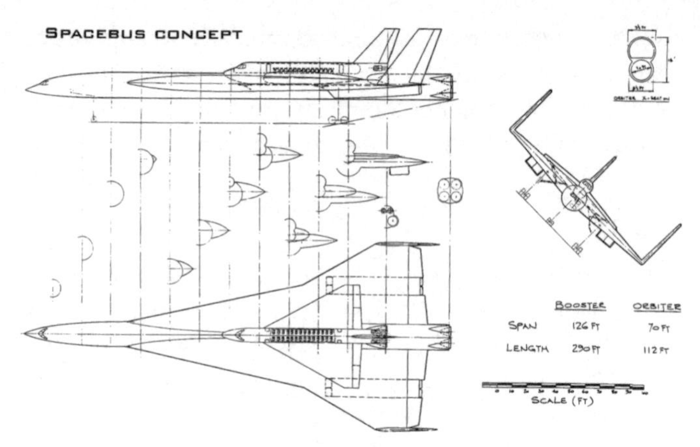

▲ **Figure 4.3 The Bristol Spaceplanes *Spacebus*: a large mature spaceplane.**

is feasible with more or less existing technology, Figure 4.3 shows the Bristol Spaceplanes *Spacebus*, which has a take-off weight comparable to that of a Boeing 747. The design is broadly similar to some of the European 1960s spaceplane projects, especially the Dassault Aerospace Transporter, except that it is larger. Its technology is such that a prototype could be built at low programme risk in a timescale comparable to that of an advanced aeroplane. As discussed in Chapter 3, spaceplanes along these general lines were considered feasible in the 1960s and there have been major technical advances since then.

This large mature spaceplane has the unavoidable changes mentioned previously. It has two stages (a Carrier Aeroplane and an Orbiter), uses some hydrogen fuel, uses rocket engines, has reaction controls and thermal protection, and has a shape suitable for high speed. It also has some optional design features, mainly in the interests of safety, the reasons for which are given in Chapter 12. The Carrier Aeroplane takes off from a conventional runway and uses jet engines to accelerate to Mach 4 (twice the speed of Concorde), which is about the upper limit with existing technology. It then uses rocket engines to accelerate to Mach 6. The Orbiter is then released and carries on to orbit while the Carrier Aeroplane flies back to base. The Orbiter as shown can carry 50 people. A cargo version can launch a 5-tonne satellite. When in orbit, it transfers people to a space station or space hotel, or releases a satellite. It may pick up a return payload before returning to Earth. Both stages are piloted.

Taking these design changes into account, the cost per flight of a 747-sized mature spaceplane will be about three times greater than that of a 747 itself. The derivation of this number starts from 747 costs broken down into fuel, amortization, insurance, crew, maintenance, and airport charges. These are then scaled using simple but conservative factors, resulting in the times three total.

The most important factors behind this increase in cost compared to the 747 are having two stages, using hydrogen fuel, and additional complexity. A more important factor affecting the cost per *seat* is that the high propellant weight and the resulting need for two stages will reduce the number of passengers that can be carried by a factor of about ten (50 seats in the spaceplane compared with 500 in the 747). This is in line with earlier comments on the penalty of using two stages in terms of lower payload. The cost per seat will then be about 30 times greater (three times the cost per flight multiplied by ten times fewer seats) than the $1000 mentioned previously for a seat in an airliner on a long-distance flight, or about $30,000 for the sake of a round number. (All these cost numbers are of course approximate.) This is a lot of money but is still about 1000 times less than the present cost of sending someone to orbit. A few days in a space hotel will cost approximately twice the transportation cost, or about $60,000 when all systems are mature.

I must emphasize that this low cost (compared to that of travelling to space at the time of writing) depends on spaceplanes achieving design and operational maturities comparable with those of a modern airliner. In particular, spaceplanes must have a long life with reasonable maintenance costs and be capable of several flights per day to space.

In summary, mature spaceplanes will enable access to space to become comparable in difficulty to present-day access to Antarctica. Space travel will not become an 'everyday' occupation, at least not using existing technology, but transportation cost should no longer be a major barrier to scientific or commercial activities. With mature developments of more or less existing technology, we can achieve a 1000 times reduction in the cost of sending someone to space. One thousand times! This would lead to nothing less than a revolution in spaceflight, as discussed in the next chapter.

5

Revolution

Before considering further the revolution in spaceflight, let us consider two previous revolutions in transportation, triggered respectively by the invention of the steam locomotive and the aeroplane (see Figure 5.1).

(a)

(b)

▲ **Figure 5.1 Harbingers of revolution: Stephenson's Rocket of 1829 and the Wright Flyer of 1903 (both shown here as replicas).**

Before the steam locomotive, the maximum speed of land transport was that of the horse. Travel between cities was slow and expensive, and most people hardly moved beyond the next village. The steam locomotive allowed speed to be greatly increased and the cost reduced. It led to an explosive growth in land transport. Stephenson's Rocket of 1829 is generally credited as being the progenitor of the 'modern' steam locomotive. It was the first to combine innovative design features that were thereafter adopted almost universally (multi-tube boiler, blast pipe, direct connection between cylinders and wheels).

Before the Wright brothers flew the first practical aeroplane in 1903, the only way to fly was in a balloon. These were used for several niche purposes such as passenger experience flights, artillery spotting, and atmospheric research. However, balloons cannot fly into wind and therefore cannot be used for an airline service. The Wright brothers showed that flight into wind at reasonable speed was a practical proposition, and this led to an explosive growth in aeronautics.

In a similar way, expendable launchers are far too expensive and unsafe for an airline service to orbit. However, as we have seen in Chapter 4, spaceplanes can provide the low costs needed for large-scale access to space. They can also provide greatly improved safety because, just as aeroplanes are inherently far safer for passengers than ballistic missiles, the same applies to their spacefaring successors, and spaceplanes will be far safer than expendable launchers (as discussed in more detail in Chapter 11). Thus, the first orbital spaceplane is likely to trigger a revolution in spaceflight comparable to the revolutions in land and air transport triggered by Stephenson's Rocket and the Wright Flyer.

As soon as the first successful orbital spaceplane enters service, it will be able to undercut any expendable launcher of comparable payload. Lower costs and improved safety will open up new markets and increase traffic levels, which in turn will release funding to enlarge and mature the design. This will further reduce costs and increase traffic levels, thereby releasing even more funding for product improvement. The result will be a beneficial downward cost spiral until the lower limit of spaceplanes using mature developments of existing technology is approached. This improvement will be similar to that of steam locomotives and aeroplanes following the pioneering work of Stephenson and the Wright brothers. Within a few decades of their developments, land and air transport had changed beyond recognition. For spaceplanes, the time taken to progress down this beneficial downward cost spiral will depend on how rapidly the new markets develop to provide the required funding, which is discussed in the next chapter.

One big difference between then and now is that Stephenson and the Wright brothers were working at the limits of the technology of the day, whereas, as was discussed in Chapter 3, spaceplanes were widely considered feasible in the 1960s. A second difference is that Stephenson and the Wright brothers had access to sufficient funds to carry out their pioneering developments, whereas an orbital spaceplane is probably at present beyond the means of individual entrepreneurs. A third difference is that developments in land and air transportation at the time of Stephenson and the Wright brothers were generally led by the private sector with governments providing some support. Space policy has been dominated by large government agencies.

The first orbital spaceplane is therefore the key to the new space age. The revolution will be one of perception as much as one of engineering. Spaceplanes will introduce an aviation way of operating, which in turn will transform the image of spaceflight from exotic to routine. Space will lose the 'exceptionalism' that has enabled extraordinary practices like throwing away the vehicle after each journey to persist for so long. Space operations will become subject to the same sorts of checks and balances that affect terrestrial transport. Space will come down to Earth, so to speak. The change from expendable to reusable launchers will be seen to be as profound as the change from balloons to aeroplanes. The perception of low-cost access to space using existing technology will change from 'far too good to be true' to 'why wasn't it done years ago?'

The new space age

Developing spaceplanes to their full potential depends on large new markets emerging to generate the necessary funding and enthusiasm. This chapter considers each major space market, existing or projected, to explore how it might develop given the benefits of spaceplanes. The aim is to see if demand is likely to be high enough to require a long production run of spaceplanes which, like the proverbial chicken and egg, is needed to achieve the low costs in the first place.

The discussion here is limited to what is likely to happen with mature developments of existing technology. More advanced technology is considered in Chapter 10.

Satellites

Looking at various transport systems, there is a correlation between the cost of using the vehicle in question and the value of the payload that it carries. You would not send coal by air, for example, except in emergencies like the Berlin Airlift of 1948–9. Conversely, you would not carry VIPs in coal wagons, except in a comparable emergency. The high cost of launching satellites partly explains why they are so expensive. A large communications satellite, for example, costs typically $500 million dollars. A large land-based microwave tower costs typically 1000 times less. (Satellites cover a far larger area, which is why they can compete despite their high cost.)

More specifically, the high cost of launching satellites (at present in the region of $5 million to $15 million per tonne to low orbit) forces a very lightweight construction, which is expensive to achieve. More importantly, once launched, satellites are on their own. The cost of sending up a mechanic to service or repair them is prohibitive, with rare exceptions like the Hubble space telescope, which was so expensive and useful that repairing it in space justified several flights of the Space Shuttle. Consequently, satellites have to be designed to work reliably and untended for several years, which again is very expensive to achieve. Difficulty of access also leads to the need for remote control. Moreover, the cost of sending up mechanics to assemble large spacecraft in orbit has limited the size of satellites to the capacity of a large launch vehicle (with a few exceptions like the International Space Station).

However, when spaceplanes introduce far lower launch costs, these limitations will be removed and we can expect satellites to be more like their land-based counterparts. They will be larger than at present and more capable, and will cost far less. They will be regularly serviced, either in orbit or in terrestrial workshops, having been returned for that purpose. Assembly of large spacecraft in orbit will become routine. We can therefore expect significant advances in those applications that satellites are so good at providing, such as global communications, weather forecasting and earth observation. Among the likely benefits are high-speed broadband links to rural areas where the low population density makes cable uneconomic.

These benefits will apply less to long-distance robotic probes that will travel beyond the point of possible repair by mechanics. However, these can still be made larger and can be checked out in space before being sent to distant parts. A higher level of redundancy will be affordable, so that a single failure is less likely to cause the loss of the mission. Lower costs will enable more to be built, which in turn will make available standard components of proven reliability.

Space stations

Space stations are satellites equipped for human habitation. To date, they have cost even more per unit weight than unmanned satellites. The International Space Station has so far cost more than $100 billion, which is an average of more than $5 million per day per occupant. It is interesting to compare space stations with airliners. Both have a pressurized cabin for human occupation. However, space stations do not need wings, tails, large engines, or landing gear. They can hardly get lost and do not have to land at night in bad weather. They do, however, need enhanced systems for life support and power provision. They also need systems for attitude and orbit control. The low cost of access enabled by spaceplanes will result in space stations being built in significant numbers. When this happens, the cost changes due to these different requirements should broadly balance out, and their production cost should then approach that of airliners of comparable capacity, which is more than 100 times less than the cost of the ISS.

Space science

There is an interesting comparison between space-based and terrestrial telescopes. The forthcoming James Webb space telescope will have a diameter of 6.5 m and will cost about $8 billion. By contrast, the forthcoming European Extremely Large Telescope (EELT), being built in Chile, will have a diameter of 39.3 m and will cost about $1.2 billion. The information gained by a telescope is largely proportional to the area of the mirror or lens, which is proportional to the square of the diameter. Thus, ratioing directly by the square of the diameter and inversely by cost, the EELT offers some 200 times more 'bang per buck' than the James Webb telescope. However, the James Webb telescope will make up for this by using the clear view from space. (The two telescopes operate at different wavelengths and are for different purposes, so the comparison is not precise.)

But why the big difference in cost? The only fundamental difference is the cost of transportation. This may be a small part of the total cost of the James Webb Telescope but, as discussed in the 'Satellites' section, it forces a lightweight construction, the need for exceptional reliability, the need for remote control, and limits the size to the capacity of a single large launcher.

Low-cost access to space will enable space-based instruments to become as large as land-based ones (or possibly even larger) and not much more expensive. Ingenious new designs might be possible, making use of the lack of effective gravity and unlimited room. Such developments will lead to nothing less than a new golden age of astronomy.

These new large instruments, and others, can be pointed downwards which, combined with lower operating costs, will enable Earth, atmospheric and environmental sciences to be greatly enhanced.

Solar power

Large satellites for collecting solar power and transmitting it to Earth have been the subject of much study. Their potential is vast indeed. The energy from the Sun that reaches the Earth over the course of just three days is equal to the energy in the fossil fuels needed to keep the human race supplied with power for 100 years at the present rate of consumption. A satellite of just 155 miles (250 km) in diameter could supply all of our present energy needs, assuming 10% overall efficiency, with a very low carbon footprint. Solar panels in orbit are always in sunlight, they can always point directly at the Sun, and the radiation from the Sun is not reduced by atmospheric absorption. As a result, the energy intake for cells in orbit is on average about ten times greater than that of terrestrial ones. However, the solar power satellites have to be launched into orbit and the power that they gather has to be transmitted back to Earth. Many engineering problems remain unsolved and it remains unclear if and when solar power will become commercially competitive. The high cost of transport to space has so far prevented even small pilot schemes. Not one light bulb on Earth has yet been lit using power from space. Spaceplanes will enable the construction of research satellites large enough to explore the feasibility of collecting solar power in space for use on Earth. It is clearly prudent to do research on such future energy options as soon as the cost permits.

Manufacturing

Factories in orbit will provide near-vacuum, near-zero gravity, and low-cost power. There have been numerous proposals for manufacturing products under these conditions. The cost of transportation has so far precluded all but very small pilot schemes, although significant research has been carried out. Spaceplanes will remove a large part of the cost barrier. Improved protein crystals and semiconductor wafers are possible candidates for early manufacture. Another possible product is costume jewellery. Some orbiting blacksmith or alchemist is sure to find a metallic mixture or crystal of striking appearance that cannot be made easily on Earth, where the convection induced by gravity distorts the solidification and crystallization processes. It is too early for a reliable prediction about the likely scale of the demand for space-manufactured products.

Lunar base

A manned lunar base has long been a goal of planetary scientists. This will require regular return transport. Soon after spaceplanes have provided reliable and economical access to Earth orbit, it is likely that two types of specialist vehicle will be built for transport from Earth orbit to and from the Moon. A space tug will provide transport between Earth orbit and lunar orbit, and a lunar lander will carry payloads between lunar orbit and the lunar surface. In this way, there will be three stages to a voyage to the Moon – Earth surface to a depot in orbit, Earth orbit to another depot in lunar orbit, and lunar orbit to lunar surface. These vehicles will be similar in concept to those used for the Apollo lunar landing programme but will be larger.

The main difference from the Apollo programme is that these vehicles will be reusable. They will be maintained and refuelled in space. Compared with the Apollo spacecraft, the main development needed for reusability is long-life rocket engines. Many of the transport missions will be robotic, to save weight and increase payload, but the key facility for enabling low-cost operations is for humans to be able to intervene when required. This is very expensive at present but will become affordable with spaceplanes and reusable space tugs. As with transport to and from Earth orbit, the culture will change from one akin to operating ballistic missiles to one akin to an airline.

The lunar base itself could use modules developed for space stations in Earth orbit. A very similar space tug would be used for transport between low Earth orbit and the higher orbits used by communications and other satellites.

Regarding the potential cost of lunar operations, a simple rule of thumb for long-distance air travel is that the total cost of a journey is about three times that of the fuel used. Broadly the same applies to your car, depending on how you account for depreciation. Spaceplanes can be expected to follow this rule, at least approximately, when their design is mature and they are in service in large numbers. Much the same rule should apply to lunar tugs and landers when they are similarly developed. Using tugs and landers as described above, the amount of fuel required per unit payload for travel from the Earth's surface to that of the Moon and back is roughly five times that required to reach Earth orbit. (This includes the fuel needed to carry fuel itself to depots in Earth and Moon orbits for refuelling the space tugs and lunar landers.) So the cost of sending someone on a return trip to the Moon should eventually approach about five times that of a trip to orbit in a mature spaceplane, which works out at approximately $150,000. For cargo, this would be about $1 million per tonne, which is about ten times less than the present cost of transport to Earth orbit.

Space exploration

Low-cost access to orbit will greatly speed up space exploration. Within a decade or two of the first spaceplane entering service, we can expect to have landed probes on most bodies of interest in the Solar System and returned samples from many of them. Routine human visits will probably be confined to the Moon and other near objects until more advanced technology is developed, as discussed in Chapter 10. Using chemical rockets, the time taken for a voyage to Mars, for example, is impracticably long, except perhaps for one or two pioneering exploratory visits. We also need to find answers to the problems of decontamination and shielding against cosmic radiation, as discussed later in Chapter 11.

There is ongoing debate about whether space exploration is better done by robots or humans. On the one hand, humans are better explorers than robots. They are more flexible and have far better pattern recognition. On the other hand, they are more expensive to transport and maintain. It is difficult to make precise comparisons of efficiency, but like-for-like trials have been carried out in simulated exploration on Earth. The conclusion is that humans are between 10 and 100 times more productive per unit of time than future terrestrially controlled explorers.

If we consider the search for life on Mars as an example, as of August 2012 there had been 50 attempts to send spacecraft to Mars, the first being in 1960. These have included three successful landers and four successful rovers, some of which have worked well for several years. We still don't know for sure whether life exists or ever has. A small team of human explorers could probably answer this question in a week or so.

With reduced cost of access, much of the politics will be removed from sending humans to space and the human/robot mix for a given mission will be decided mainly on best value for money.

Mining asteroids

There have been several proposals for mining the asteroids for rare metals and, as mentioned in the Introduction and Chapter 3, a new company (Planetary Resources) has been set up to do just that. Some research suggests that all the gold, cobalt, iron, manganese, molybdenum, nickel, osmium, palladium, platinum, rhenium, rhodium, ruthenium and tungsten mined from the Earth's crust, and that is essential for economic and technological progress, came originally from the rain of asteroids that hit the Earth after the crust cooled. If this is true, asteroid mining has been taking place for thousands of years!

With mature transportation systems, the cost of a return voyage to an asteroid will be approximately the same as to the Moon, at around $1 million per tonne, depending on how near the asteroid in question is. As a very rough rule of thumb, if the transportation cost is less than 10% of the price of a particular metal, mining of that metal on asteroids might be commercially viable, assuming that the weight of the ore returned is not much more than that of the metal eventually extracted. This is of course a big assumption, so this rule of thumb gives no more than a preliminary indication. Ten times the transportation cost when all systems are mature is around $10 million per tonne, or $10,000 per kilogram.

Table 6.1 shows the rough value of some selected metals, listed in descending price per kilogram. Some of these are worth more than the ten times the transportation cost threshold; others less. These prices are of course volatile and the estimate of transportation cost is tentative, so this table gives only an initial indication.

A big advantage of asteroid mining will be the reduced terrestrial pollution due to mining.

With these assumptions, it seems that mining asteroids will be viable for some of the more valuable metals when

Table 6.1 Value of selected metals

Metal	Price	
	$/kg	
Gold	55,000	
Platinum	48,000	
Rhodium	42,000	Price more than ten times transportation cost
Palladium	19,000	
Osmium	14,000	
Rhenium	5000	
Ruthenium	4048	
Cobalt	30	
Molybdenum	30	Price less than ten times transportation cost
Nickel	15	
Manganese	3	
Tungsten	0.5	

the reusable space transportation systems are mature, if ores of reasonable quality can be found.

There is, however, some way to go. The six Apollo flights that landed on the Moon returned a total of 382 kg (842 lb) of lunar soil and rock samples between them. The total cost of Apollo was some $150 billion at 2013 prices. So each kilogram of lunar sample cost on average around $370 million, which is about 7000 times more than the price of gold. More would have been brought back if that had been the primary aim of Apollo, but these figures give some indication that new space transportation will be needed for commercial success.

Extraterrestrial intelligence

Larger space observatories will greatly increase the sensitivity of the instruments used in the search for life elsewhere in the universe. Returning samples from all bodies of interest in the Solar System will enhance our understanding of what conditions are needed for life to evolve. These developments will add insight to our search for whether or not we are alone. If this is not enough to settle the question once and for all, then the longer-term developments discussed in Chapter 10 should do so.

Space tourism

With affordable space travel, space stations could be equipped to house tourists. Many ordinary people would love to take a visit to space as soon as it becomes safe and affordable. Most astronauts to date have described the experience as transforming. Dennis Tito became the first space tourist in 2001, when he paid some $20 million for a visit to the ISS. Tito was unusually well qualified but still had to undergo six months of intensive astronaut training in Russia. As of October 2012, six other private citizens have followed Tito to the ISS, paying about $30 million each. No other holiday even approaches this cost, which is an indication of the 'magic pull' of space. Also in October 2012, Sarah Brightman, the British singer, announced that she had booked to become the eighth citizen to go to the ISS.

The main attractions are the views of the Earth, playing around in zero-g, and the clear views of various celestial bodies. Spaceplanes should enable the cost to be reduced so that middle-income people who are prepared to save could afford the holiday of a lifetime. In this section, we will consider the possible size of the market for space tourism, while Chapter 7 describes the experience in greater detail.

To give a very tentative indication of the possible size of the market, let us assume that the world's industrialized population is 1 billion people and that that just 1% of these would be prepared to pay a few tens of thousands of dollars for a visit to a space hotel once in their lifetime. These assumptions lead to an initial demand of 10 million people. Transporting these would require a fleet of about thirty 50-seat spaceplanes each making two flights per day for ten years. This preliminary but probably conservative estimate indicates that space tourism alone is likely to provide the incentive and enthusiasm needed to develop spaceplanes to their full potential. To move down the beneficial downward cost spiral rapidly requires space tourism to take off like a craze and become the 'must-do' holiday. Crazes are notoriously difficult to predict, but space tourism seems to have the essential ingredients.

This estimate of 1 million space tourists per year makes an interesting comparison with the number of people who have been to space in the 51 years of human spaceflight – a total of 528 as of 31 June 2012. The expression 'game-changer' is a tame cliché to describe the transformation that spaceplanes could soon enable.

In terms of the likelihood of providing large markets for spaceplanes, space tourism is the most promising discussed so far in this chapter.

Long-range transport

As mentioned in Chapter 3, if a spaceplane accelerates to just short of satellite speed, it can then glide halfway around the Earth, offering the potential for very fast air travel. A flight from Sydney to London, for example, would take about 75 minutes' flying time. However, a visit to space would be, for most, a once-in-a-lifetime experience for which people would pay more than for a fast flight halfway round the world. Thus, for commercial success, fast long-distance transport will require a more economical vehicle than will space tourism.

Long-range transport spaceplanes would also have to meet more demanding noise and environmental requirements because they would have to take off close to big cities whereas, if necessary, space tourists could fly from a few airports well away from population centres. Moreover, avoiding sonic bangs over land will be easier for space tourism launches and re-entries than for suborbital airliner flights, because of the greater freedom for selecting flight paths.

The long-range suborbital airliner would therefore have to be more environmentally friendly than a space tourism spaceplane as well as more economical, and hence of more advanced design. It therefore seems likely that, if it happens at all, fast long-distance suborbital air travel will take place after large-scale space tourism.

Overview

This snapshot of the new space age suggests that when spaceplanes enable access to space comparable in difficulty to access to Antarctica, then environmental and astronomical research will be transformed. Space tourism is likely to become the first commercial market for spaceplanes that is big enough to generate the funding needed for their rapid development to full potential. Manufacturing in space will happen, but the demand for space products is uncertain. Solar power generation has vast potential but faces stiff engineering and economic challenges. Long-range suborbital airliners for high-speed transport will need to be more advanced than orbital ones for space tourism. Mining asteroids should become commercially viable if high-quality ores can be found.

Other uses of space will clearly benefit from the reduced cost of access, but the breakthroughs are likely to be in science and space tourism.

Consequently, it is very likely that the demand for spaceplanes will be high enough to ensure rapid development from operational prototype to approaching airliner maturity.

7

Your space tourism experience

This chapter assumes that you have saved for the holiday of a lifetime and describes what your visit to a space hotel will be like.

Preparation

Much of the preparation will be like that for any other holiday, such as finding the time, getting the best deal, finding somewhere to park the car at the airport, securing the house, and arranging for someone to water the plants and look after the dog.

There will, however, be a health check. Astronauts to date have had to be very fit and healthy. This is because only the most basic medical facilities have been available in manned spacecraft, and evacuation has been difficult or not possible. Moreover, with taxpayers spending millions of dollars per day per astronaut, it is highly undesirable for an experiment to be lost because the experimenter becomes sick.

These factors will apply far less to space tourists. Spaceplanes can be used as affordable space ambulances, and space hotels will be equipped with medical facilities. However, the tour operators, safety authorities and insurance companies can be expected to impose moderately severe health checks during the pioneering days of passenger spaceflight. Research so far indicates that the most important test will be a ride in a centrifuge to check that your heart and circulation can cope with the g forces of launch and re-entry.

These checks will be relaxed as soon as confidence is gained from the experience of ordinary people coping with spaceflight. Some new problems may emerge while others may turn out to be less severe than expected. After a few years, most people fit for an active holiday on Earth will probably be able to visit a space hotel. People who are less fit may have to wait a few more years until additional facilities are available for their comfort and safety.

There will also be a training programme, part compulsory and part optional. The compulsory training will include the use of emergency equipment. This will include a pressure suit that will probably be needed for the flight to the space hotel, and the emergency survival equipment in the hotel itself. The pressure suit is necessary in case of loss of cabin pressure. The emergency oxygen fitted in airliners would be of little help in the vacuum of space. External pressure is needed to prevent blood from boiling and to enable breathing. The pressure suit will be far simpler than that worn by astronauts or even by high-flying pilots. All it has to do is to keep you alive for a few minutes until the spaceplane can descend back into the atmosphere. There are no requirements for mobility when under pressure, or for sophisticated thermal control, or for multi-layer protection, which drive the complexity of existing pressure suits.

The tourist suit will be loose fitting and reasonably comfortable to wear, with a visor that closes automatically to seal the suit if cabin pressure is lost. An air supply will then provide the pressure and oxygen required for life support, and the inflated suit will be nearly rigid.

There will be an emergency suit in the hotel which will be even simpler, consisting of an airtight fabric garment, stored in emergency lockers, that you zip yourself into and then await rescue. It will form a sphere of about 1 meter in diameter when under pressure. NASA has already tested such 'rescue balls', designed for possible emergency use on the Space Shuttle.

The optional part of the training will last a few days and will consist of briefings on what to expect and how to make the most of your time in space. Rides will be available to get your body used to the unfamiliar g forces that you will experience. Some of these rides will be like those in theme parks; others will be in aeroplanes flying parabolic trajectories to provide periods of low gravity. The latter are available commercially now in airliners with most of the seats removed to allow space for low-g gymnastics. These provide up to 20 seconds of low-g at a time. Supersonic aeroplanes could give about one minute, and suborbital spaceplanes two minutes and more.

One problem not yet completely solved is space sickness. This is related to ordinary travel sickness but is sufficiently different to need special research. It comes on soon after first reaching 'zero-g', and affects a significant proportion of astronauts. Most recover after a few hours. Several medications have been developed, and these are effective for most people. One purpose of the theme park and aeroplane rides will be to find the best medication for you, or indeed to see if you need medication at all.

The outbound flight

Your seat in the spaceplane will be quite similar to an airliner seat. You will, however, be wearing the emergency pressure suit. Take-off and initial climb will again be just like in an airliner. You will notice the difference when the Orbiter stage of the spaceplane, in which you are seated, separates from the Carrier Aeroplane. You will see the latter disappearing beneath you. Then, when the rocket engines are set to full throttle, you will hear a louder noise, feel considerable vibration, and feel a strong push on your back as the acceleration increases.

The higher the allowable peak acceleration during the ascent to orbit, the less propellant is needed. The acceleration will probably be set at the maximum that most people turn out to be comfortable with, probably around 2 g.

As you leave the effective atmosphere, you will see the sky turning dark. When your eyes have adapted, you will see bright stars even in daytime. At this point you will be able to see the ground for a distance of several hundred kilometres, and the curvature of the Earth.

After about six minutes of acceleration in the Orbiter, the rocket engines will stop and the noise and vibration will all but vanish. You will feel weightless for the first time. Soon afterwards, the hotel will come into view and, if you are sitting at an appropriate window, you will be able to watch the docking manoeuvre. This will be a bit like a ship berthing at a pier except that it will be in three dimensions rather than two, and the dock-hands attaching lines will be replaced by robot arms.

A few minutes later you will float weightlessly down the aisle, using 'banisters' to guide you, through the cabin door and the air-lock passage and into the hotel.

The hotel

There will be two sections to the hotel, one gently rotating and the other still. The rotating section will provide a low effective g level at its rim, due to centrifugal force, probably about one-sixth g, so that eating, sleeping, keeping clean, going to the toilet, and other everyday activities are comfortable and do not require special equipment or training.

The non-rotating section will house the viewing ports and the zero-g gymnasium, sports hall and swimming pool. The view of the Earth is continuously changing as the hotel orbits every 90 minutes. There is a sunset and sunrise every orbit. Features that cannot be picked up even from high flying aircraft become apparent, such as giant eddy currents in the oceans hundreds of kilometres across, complete cyclones, geological fault lines, and meteor craters. You will be able to identify your own house through a telescope, provided the sky is not cloudy at the time and the orbit passes within viewing distance.

On the dark side of the Earth, you will see a dozen or so lightning flashes per second. The aurorae near the magnetic poles can be seen from above as rippling cones of lights.

Telescopes will enable you to see the Sun, Moon, planets, stars and galaxies far more clearly than from on Earth, where the atmosphere distorts the image.

In the zero-g gym you will be able to fit wings to your arms and tails to your feet and literally fly like a bird, looking like a medieval or renaissance birdman. Your muscles will be strong enough to propel you through the air in the absence of apparent gravity. Your motion will, however, be different from that of a bird on Earth because of this lack of gravity. If you stop flapping or twisting your wings or tail, you will carry on in a straight line (relative to the gymnasium), rather than falling towards the ground.

An interesting experiment will be to see how real birds cope with this situation – especially penguins! After some practice, you may end up being able to fly better than eagles. Human flying races will be one of the many new sports made possible by weightlessness.

Zero-g swimming will be another new experience. The pool will consist of a large cylindrical drum rotating slowly about its main axis, like a huge washing machine drum. It will contain enough water to provide a metre or so depth around the rim, which is where the water will go due to the rotation. Perhaps surprisingly, you will sink to the same depth as in a pool on Earth, with just your head showing. This is because your apparent weight and the hydrostatic pressure that provides the buoyancy force both decrease by the same fraction as the apparent g is reduced.

The absolute force pulling you down in the pool will be far less than it would be on Earth (assuming that the drum is rotating slowly) and you will be able to kick off from the 'bottom' and land on the water at the 'ceiling'. Your trajectory will be straight relative to the non-rotating hotel, but curved relative to the drum, because of the rotation. With small flippers attached to your arms, you will be able to take off and fly like a flying fish.

There will be optional lectures on geography, Earth science, and astronomy, supported by superb views of the objects in question.

The return flight

You will return to Earth in the type of vehicle that you flew up in. The return flight will be similar to the ascent, except that there will be no Carrier Aeroplane and the acceleration will be replaced by deceleration. During the peak of re-entry heating, all you will see out of the window will be a hot glow.

Hopefully, like most astronauts, you will look back on your first space flight as a transforming experience.

A flight in a suborbital spaceplane will be broadly comparable, except that you will stay in the vehicle and have only a few minutes in space.

Who owns space?

The greatly increased use of space resulting from the introduction of spaceplanes will draw new attention to space law. Space law has been evolving since the earliest days of spaceflight and generally speaking has served us well. It has not held up scientific or commercial progress, and has led to more or less amicable resolution of most conflicts. This probably explains why most of us have hardly heard of space law.

Space law

Space law dates back to the late 1950s, when US intelligence agencies found out about the rapid Soviet progress towards launching the first satellite. President Eisenhower was pressed to launch a satellite ahead of them to gain the prestige of being first. He resisted this pressure largely because he wanted satellites to be free to fly over other countries. He argued that if the USA were first, the Soviets would make a big fuss and try to ban future satellites. If, however, the Soviets went first, they could hardly argue. This is indeed what happened, although Eisenhower lost short-term prestige as a consequence.

Five treaties have been negotiated and drafted under the auspices of the United Nations Committee on the Peaceful Uses of Outer Space. These are the Outer Space Treaty, the Rescue Agreement, the Liability Convention, the Registration Convention, and the Moon Treaty.

The last of these is likely to be under the most pressure when we develop low-cost access to the Moon. The Moon Treaty's provisions are not unlike those protecting the Antarctic, our only continent without an indigenous population. The Antarctic treaty seems to work reasonably well. Unfortunately, the Moon Treaty is regarded as a failed treaty because it has not been ratified by any of the leading spacefaring nations. The main source of disagreement is how to share the proceeds of exploitation. However, when it comes to matter, there will be much international pressure to ratify it, or a modified version. There are two positive signs. First, the future exploration of the Moon is likely to be an international affair. Second, the plaque left on the Moon after the first landing and signed by President Richard Nixon did say 'We Came in Peace for All Mankind'. How well this situation is resolved will be a good test of international good will.

Vehicle certification

Another area of regulation that is becoming important is vehicle certification. A new type of airliner is designed to various airworthiness codes, which have evolved over the 100 years of commercial flying. It is tested to demonstrate compliance with these codes, which includes typically 1000 test flights. The design is then awarded a Type Certificate. Individual airliners that are shown to conform to the type design then receive Certificates of Airworthiness.

There are two difficulties in applying this procedure to spaceplanes. First, the airworthiness codes for operating in space have not yet been written and there is a lack of operational experience to base them on. Preparing them will require drafts to be issued and then refined in the light of operating experience.

Second, the required testing would be prohibitively expensive for a new industry that still needs to develop a solid market base. (Developing a new aeroplane to full certification is typically ten times more expensive than building just a prototype.)

The USA, with its unique tradition of manned spaceflight, is taking a different approach. The Federal Aviation Administration is regarding pioneering spaceplanes more as launch vehicles than as airliners. Accordingly, the main concern is possible damage to people and property on the ground. Early operators will be allowed to carry passengers on an informed consent basis. Passengers will have the risks fully explained to them and then asked to sign a consent and waiver of liability form.

Another approach is to regard spaceplanes as airliners and, in order to enable the industry to get started at an affordable cost, bring in a sort of 'certification lite' for early operations. After some early test flights, prototypes would be allowed to carry passengers on an informed consent basis, under restricted operating limitations. For example, flights might be restricted to good weather in daytime from specific airfields. They might be flown by accredited test pilots only. Each flight might be regarded as a test flight, with appropriate instrumentation and post-flight analysis. Under the close supervision of the authorities, insurance companies, operators and manufacturers, the operating limitations would be progressively relaxed in the light of operational experience. Eventually, full certification would be achieved and 'normal' commercial operations would take over.

The question of spaceplane certification will certainly receive considerable attention over the next few years.

9

Space and the environment

This chapter considers the environmental balance sheet of large-scale space operations.

The main disadvantage will be increased atmospheric pollution. Spaceplanes climbing to space will inject pollutant into the high atmosphere. It is likely that the propellants will be liquid hydrogen and oxygen. The exhaust product will therefore be water vapour, which is non-toxic. It may, however, be persistent and the effects will need to be studied carefully.

There will be several benefits. In approximate time order, the first benefit will be the greatly reduced cost of environmental science from space, which is key to understanding human impact on the environment. There will be a rapid gain in environmental knowledge, which will help us to restrict our polluting activities in a more effective manner. The large Envisat satellite, for example, was launched in 2002. It worked well for ten years before being declared no longer operational. It led to much ground-breaking research in the following fields:

- atmospheric chemistry
- ozone depletion
- biological oceanography
- ocean temperature and colour
- wind waves
- hydrology
- agriculture and arboriculture
- natural hazards
- digital elevation modelling
- monitoring of maritime traffic

- atmospheric dispersion modelling (pollution)
- cartography
- study of snow and ice.

However, the Envisat satellite cost about $2 billion to build and launch, and very few such satellites can be afforded. However, with low-cost access by spaceplane, several large environmental satellites of this kind could be afforded. They could be optionally man-tended and could be even larger and more capable than Envisat.

The second benefit will be a boost to the hydrogen economy. Spaceplanes are likely to be the first large-scale commercial users of liquid hydrogen fuel, which in turn will make it easier for it to be adopted for aeroplanes and ground transport systems.

The third benefit, which is longer term, is that using resources from space will reduce the human pressure on our own planet. Two likely candidates for this are solar power satellites and mining asteroids.

The fourth benefit, which is also longer term, is that low-cost access to space would facilitate setting up a defence against asteroid impact, and developing climate modification technologies, should these become necessary.

The fifth benefit, less directly linked to the environment, will be new jobs and economic expansion. Space tourism in the near term, space manufacture, and maybe solar power satellites later on will create large numbers of new skilled jobs.

These disadvantage and benefits can be reliably expected. The remaining benefit is more speculative because it involves human psychology. What will be the effects on human thinking of large-scale space tourism, space art, space sport, and enhanced space science and exploration? Some indication is available from the experience of astronauts. Most have said that visiting space was a transforming experience, and that they would like to go again. They tend to return to Earth with a more global perspective than when they left: they are more conscious of the fragility of 'Spaceship Earth'. When a million and more people visit space each year, we can expect these views to spread, which should make it easier to generate global action to counter the various perils that face our home planet.

There are two historical precedents. First, Europe benefited greatly from discovering the existence of America. The resulting new knowledge and challenge to traditional thinking provided a boost for the Enlightenment. Could large-scale space exploration and tourism provide a comparable benefit to the entire planet?

Second, many people consider that the main benefit from the Apollo lunar landing programme was the famous 'Earthrise' photograph (see Figure 9.1) of the Earth appearing to rise above the Moon.

To quote from a senior NASA Apollo engineer, Henry O. Pohl, looking back on Apollo:

> ***'Probably the most significant benefit of the Apollo programme was pictures from space, allowing everyone to see the Earth for what it was – a little ball with a very thin, fragile atmosphere around it. One picture from Apollo of the whole Earth caused the entire world to start thinking about what we were doing***

▲ Figure 9.1 'Earthrise': Taken from Apollo 8 by Frank Borman in December 1968. This photograph soon became the icon of the environmental movement.

to the environment and the planet. That picture may have done more for mankind than any other single thing – just giving us a perspective of the world we live in. That's something we don't normally think about. For the first time man could see the Earth all at one time and realize it as a very small, very fragile planet, and we were really destroying it fast.'

I remember seeing 'Earthrise' for the first time and, like billions of others, starting to take Greenpeace and Friends of the Earth seriously.

There is therefore little doubt that the environmental benefits of the large-scale use of spaceplanes will far outweigh the disadvantages.

10

Space colonization

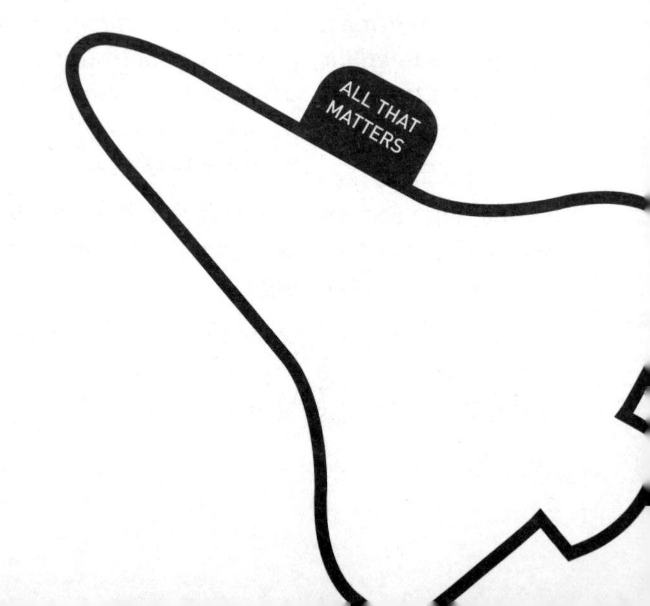

This chapter considers the prospects for colonizing space beyond the Moon, and speculates about some possible futures for the human race in the universe. There is no shortage of articles and books on the subject. Some suggest that space colonization is the next breakthrough in evolution, like the first animals coming out of the sea to colonize the land. Others suggest that space provides an answer to the 'limits to growth' on the home planet, which are already leading to increased regulation and reduced individual freedom. Others point out that a frontier, now all but gone on Earth, is needed as an outlet for restless spirits, and that massive space exploration might restore a sense of optimism to human endeavour. Yet others think that we should not explore space at all until we have better sorted ourselves out here on Earth.

Let us consider what might be possible from an engineering viewpoint. So far, we have been looking at what is likely to happen in space soon, using more or less existing technology. This means using chemical rockets, i.e. rocket engines that generate heat by the chemical reaction between two (occasionally more) substances. These are already close to their theoretical performance limits. A tonne of hydrogen and oxygen contains just so much energy. As we have seen, chemical rockets have adequate performance for return manned visits to the Moon and robotic one-way visits to the edge of the Solar System. The next target for human exploration is Mars. Using chemical rockets, the fastest practicable time for a round trip is about 250 days. This might be acceptable for a few pioneering human visits, and it might be achievable and affordable within, say, 10 to 15 years of the first orbital spaceplane entering service, assuming that we can develop lightweight shields against cosmic radiation and solve the problem of decontamination (as discussed later in Chapter 11).

However, for regular human visits to Mars and exploratory visits further afield, we need faster travel and that requires new technology. There have been many proposals for achieving higher speed, a few of which are summarized below.

Nuclear rockets

In a nuclear rocket, the heat is generated in a nuclear reactor. This is used to heat a working fluid, usually liquid hydrogen, which expands through a rocket nozzle to produce thrust. Compared with chemical rockets, nuclear ones can produce about twice the impulse (thrust multiplied by time) for a given weight of propellant, which would greatly speed up trips to Mars. Nuclear engines are heavy and their use is likely to be restricted to in-space operations.

The USA tested several types of nuclear rocket from the 1950s to the 1970s, and produced designs that were almost ready for production. Research at a lower priority continues to this day. The main problem is the risk of contamination in the event of catastrophic failure during launch. Reliable crash-proof containers are feasible, but persuading public opinion might be more difficult.

One longer-term solution might be to build the radioactive components in space or on the Moon.

Small quantities of radioactive materials, up to about 20 lb (9 kg), have already been used in some probes to provide power on trips to Mars and further afield, where the radiation from the Sun is weak and solar cells provide less power.

Electromagnetic catapults

An electromagnetic catapult uses a linear electric motor to accelerate a payload along a rail. This could be used to launch spacecraft from the Earth to orbit, from Earth orbit to more distant parts of the Solar System, or from the Moon or other bodies. Electromagnetic catapults are used to accelerate some adventure park rides and are planned for launching aircraft from the decks of the next generation of US Navy carriers. The technology is largely available for longer and faster systems.

Launching from Earth to orbit poses two problems. First, the rail would have to be long. To reach orbital velocity and allowing for drag and gravity losses, and assuming a 4-g acceleration, which is about the limit for humans over that period of time, the rail would have to be nearly 900 miles (1400 km) long. Cargo can accept higher accelerations, and a shorter rail could be used.

Second, the air resistance and heating on leaving the end of the launch rail (which would be in a tube from which the air had been evacuated) would be unfeasibly high if the tube exit were at sea level. The exit would have to be higher than Mount Everest to make much difference.

If these problems can be solved, an earthbound catapult would greatly reduce the cost of launching payloads to orbit.

A catapult in Earth orbit would avoid the problem of air loads, but the cost of installation would be high, even with low-cost launch. Every time it was used, the catapult would be propelled backwards, in reaction to the force used to accelerate the payload, so rockets would be needed to return the catapult to its correct orbit. These could use electrical power to propel the working fluid. There is virtually unlimited electrical power in space, using solar panels. Electric rockets are very efficient in terms of propellant usage but are too heavy to be used for Earth to orbit transport. They are used for keeping satellites on station and for changing orbits.

A catapult on the Moon would avoid the problem of maintaining the correct orbit. Moreover, the Moon's gravity is lower than Earth's and so the speed needed is lower. A lunar catapult to launch humans into a trajectory back to Earth would be 'only' about 100 miles (160 km) long.

Space towers and space elevators

A space tower is a structure that reaches up to space, or near to space, to provide a platform for low-cost launch. A space elevator uses a long cable to anchor a satellite that is so high that it needs a pull greater than gravity to maintain its orbit such that it remains above a given point on the equator. To reach space, an elevator would be attached to the cable.

A point on the equator is moving at about 6% of satellite speed, which is why eastbound launches from the equator are the most efficient in terms of the payload that can be carried. If you built a vertical pole at the equator, the speed at the tip would be higher. The longer the pole, the higher the speed. (In a similar way, your speed over the ground increases as you move away from the centre of a playground roundabout.) Eventually, the centrifugal force arising from the circular motion becomes significant, thereby reducing the effective weight of a mass on the top of the pole. At a certain height, the centrifugal force balances gravity (allowing for the fact that the force of gravity reduces with increasing distance from the Earth) and the mass would have zero effective weight. A satellite released here would stay put and maintain its position over the equator. This so-called geostationary orbit is used by communications satellites to cover a fixed area of the Earth. Geostationary height is 22,000 miles (36,000 km). At greater heights, the mass would need a downward pull from the pole in order to stay at the same point over the equator; hence the idea of a space anchor, to which could be attached an elevator. (A free-flying satellite below geostationary height moves faster than a mass on the pole at the same height; one above geostationary height moves slower.)

The structures of space towers and anchors are subject to the force of gravity and are not feasible with present materials. The tallest man-made structure is about 2600 ft (800 m). Materials with much higher strength-to-weight ratios are required for space towers and elevators. The required properties have been approached in the laboratory using very small samples. While space towers and elevators may be built one day, it is too early to say when.

Other new technologies

Several other techniques have been studied for increasing the speed or distance of space exploration. Humans need faster transport than cargo. So, to speed up human transport, pre-positioned depots can be set up containing supplies, fuel and habitats. In a similar way, pre-positioned fast horses were used to speed up imperial messengers in ancient times.

Setting up these depots can use slower transport such as solar-powered electric rockets, or solar sails that use the pressure of radiation from the Sun to propel spacecraft slowly but over long distances. It may be possible to use materials from planets or their moons to provide some of the supplies for these depots.

Overview

Nuclear rockets and large catapults have been studied reasonably thoroughly and are clearly feasible. Low-cost access to orbit would provide a strong impetus to the development of such new systems. In this way, we can expect spacecraft launched by electromagnetic catapults and powered by nuclear rockets to make the Solar System much more accessible. Studies are needed to establish possible timelines for visiting increasingly distant bodies.

Whether or not we will want to emigrate to distant parts is another question. What would be the point? Perhaps the base at the South Pole is a useful analogy: it is permanently occupied by a transient staff; no baby has yet been born there.

Speculation

So far, the discussion in this book has been based on historical fact, reliable studies, and informed opinion. The following comments are highly speculative, so please do not judge the credibility of the earlier part of the book by what follows in this chapter!

Travelling even further afield might involve technologies like nuclear fusion rockets, anti-matter, and zero-point energy propulsion. However, these are at present so speculative that it is too early for a useful discussion of feasibility or possible timelines.

However, we can get a feel for the eventual limitations by making some simple assumptions. Let us assume that eventually we can develop robotic space explorers that are sent out in different directions. Each of these is programmed to search for a suitable planet where it would land and replicate itself many times over to provide more explorers that then take off again and look for more planets, and so on. Assume that the average rate of advance is one quarter the speed of light. The most distant part of our galaxy is some 90,000 light years from Earth, so it would take about 360,000 years for our probes to reach the edge. This is short enough by cosmic timescales, but is a long time for us humans. And that is just our galaxy. There are hundreds of billions more.

From these simple sums, it is reasonably clear that with *Homo sapiens* in our present form and with the laws of nature as we understand them, space colonization in the sense of humans visiting the edge of our galaxy is not going to happen. So either *Homo sapiens* or the laws of nature as we understand them have to change dramatically.

Of these two, It would seem that *Homo sapiens* is the less unlikely to change in the foreseeable future. The transhumanist movement, for example, talks about turning ourselves into 'posthumans' with greatly enhanced abilities. These might include the ability to hibernate indefinitely in the aforementioned robotic explorers, or even the ability to manipulate the universe itself. Others doubt that this can ever happen or, if it does, that it will be thousands of years in the future. These prospects inspire hope, fear or scepticism, depending on whether one inclines towards optimism, pessimism or realism.

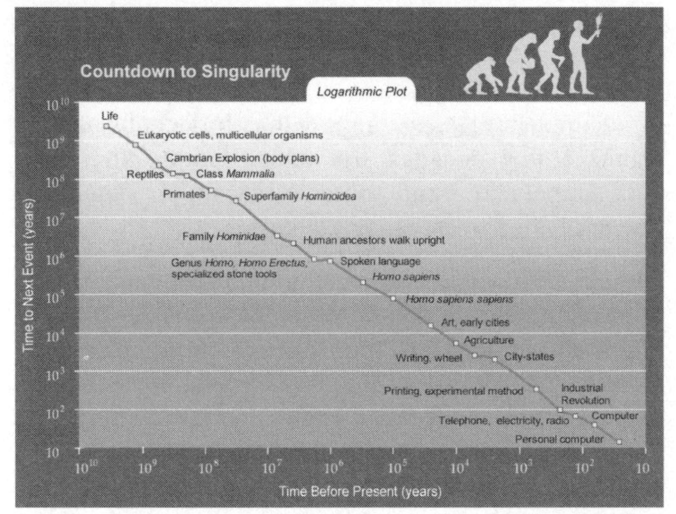

▲ **Figure 10.1 Technology development timeline on a log scale. (Taken from *The Singularity is Near* by Ray Kurtzweil, Viking, 2005)**

The one thing that can be said for sure is that the rate of achieving major breakthroughs has accelerated exponentially over time. This is shown in Figure 10.1.

In Figure 10.1, breakthroughs of roughly comparable significance are plotted, showing how long ago they happened. Approximately speaking, they fall on a straight line on a log scale, which indicates exponential development. The time to the next breakthrough is decreasing very rapidly. For example, moving from the Industrial Revolution to the personal computer happened 1000 times faster than from *Homo sapiens* to agriculture. Certainly, the runaway growth of electronic and information technologies in recent years is without precedent. Ray Kurtzweil, who originated this chart, predicts that the next breakthrough will happen when computer capability overtakes that of the human brain in all important aspects, and that this could happen within a few decades. He calls this turning point 'The Singularity', and there is already a Singularity University in Silicon Valley, California.

There is no obvious end in sight to this exponential rate of development. It therefore seems plausible that the timescales for long-distance space colonization could be comparable to those of the development of the human race itself, and that these could interact in some way. Perhaps it is not too fanciful to speculate that we will achieve our final destiny, whatever that may be, in some as yet unimaginable confluence of matter, space, time, and posthuman evolution.

But the immediate action is to build the first orbital spaceplane.

Safety

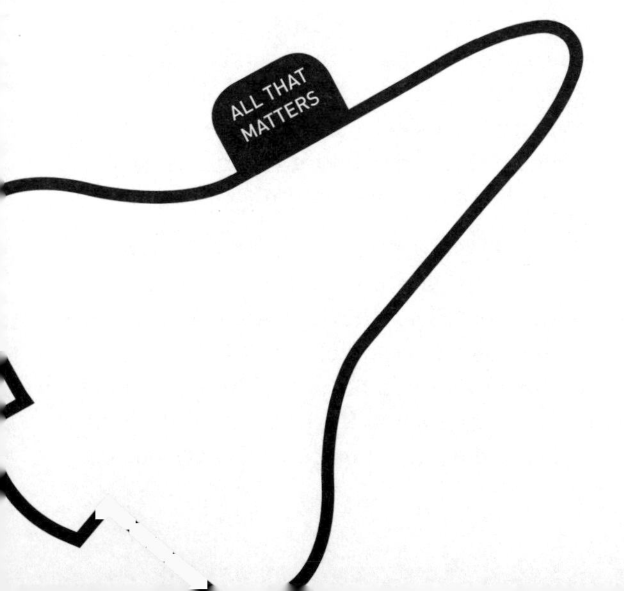

Before considering the roadmap to the first orbital spaceplane, this chapter discusses safety, which will become more important as greater numbers of people visit space, and which will have a large effect on the design of spaceplanes.

Spaceflight accidents

There have been 18 astronaut fatalities during launch or re-entry, 14 in the US space programme and 4 in the Soviet (as of June 2012). This is about 1 in 30 of the 528 people who have visited space. There is an interesting comparison with climbing Mount Everest. In the 10 years to the end of 2010, there were 54 fatalities and 3938 climbs to the summit, which gives an average of 1 fatality per 73 summits. There were fatalities among people who did not reach the summit, so this average is not the complete story. But this one in 70 (for the sake of a round number) is good enough for our purposes. Thus, manned spaceflight to date has been somewhat riskier than climbing Everest. Incidentally, the cost of joining an expedition to climb Everest is around $60,000, which, as discussed earlier in Chapter 4, is comparable to the future cost of a stay in a space hotel when all systems are mature.

Most of these spaceflight accidents can be put down to the one-off failures that are inevitable in a pioneering activity involving expendable vehicles operating at great speed in an unfamiliar environment. There have been no Soviet/Russian fatalities since 1971, and if a new launcher were to be built along the general lines of the Space Shuttle it would no doubt be considerably safer.

However, the use of large, complex, throwaway components limits the safety that can be achieved. Vehicles off the production line cannot be tested in flight before being used, because they can fly once only. Many safety-critical components therefore have to work correctly the first time. By contrast, airliners off the production line usually make two or three verification flights at the manufacturer's airfield, followed by a handover flight to the customer airline, which will generally put the aircraft directly into service.

The high cost of launches means that very few can be afforded, and there is no way that expendable launchers can approach airliner maturity. Moreover, the inherently high cost per flight of expendable vehicles severely restricts the number of test flights that can be afforded. For example, the first powered flight of the Space Shuttle went all the way to orbit. Each system had to work right the first time, close to its limits. The decision to do this was not taken lightly and, due to great skill and dedication, it worked. However, some luck was involved. The first re-entry showed that the angle of the flying controls needed to stabilize the vehicle was twice that predicted, and not far short of their maximum angle. Had the error been much greater, the vehicle would probably have crashed. This type of problem would normally be observed in an incremental flight-test programme long before it became serious.

Such a leap into the unknown would be unthinkable with a new aeroplane. For example, Concorde (which had a maximum speed of Mach 2, which is about 12 times slower than the Space Shuttle) made 69 flights before reaching even supersonic speed, and a new type of airliner requires some 1000 test flights before being allowed to carry passengers. Aeroplane flight testing follows an incremental programme. Each flight 'pushes the envelope' a little. It is a little faster, or slower, or higher than previous flights, or tests a system nearer to its design limit. If something does not work as planned, there is a proven flight envelope to fall back into.

Expendability explains why the Space Shuttle was such a risky project, with 2 fatal accidents in just 135 launches over a 30-year period. The safety of the crew depended on large and complex single-use components. Both fatal accidents were caused by failures in these components causing damage on the Orbiter. Each time the Space Shuttle flew, a crew of seven was in effect flight-testing a ballistic missile. Expendability is thus the major cause of the high risk of spaceflight to date.

The most reliable expendable launchers are approaching 100 flights between major failures. By using a separate capsule for crew and passenger escape in case of major failure in the launcher, it is likely that 1000 flights per fatal accident could be achieved with an expendable system.

This is considerably better than has been achieved so far, but is still much more risky than airline flying. In the five years up to 2011, the average number of airline passengers killed per year was about 1000. In 2011, there were 2.8 billion passenger flights. Therefore, your chance of being killed each time you make a flight is about 1 in 3 million, which is some 3,000 times safer than spaceflight would be using new and more reliable expendable launchers. Airliners are the safest flying machines yet invented.

Spaceplane safety

Spaceplanes will be far safer than expendable launchers because they are aeroplanes in engineering essentials. An adequate number of test flights can be afforded, and lower cost will enable far more flights to take place, which in turn will enable airliner maturity to be approached. Much of the technology can be transferred from that used in airliners, with all the resulting background of testing and operations.

However, spaceplanes will be exposed to risks that do not affect airliners. They have to operate in the vacuum of space, so the integrity of the pressure cabin is even more important than it is in an airliner. They are exposed to the heat of re-entry. Rocket propulsion is potentially more hazardous than jet propulsion because both the fuel and oxidizer have to be carried on board. (With airliners, the oxidizer is the oxygen from the atmosphere.) These chemicals tend to react violently when they mix, which means that leaks from propellant tanks, pipes, valves, pumps and engines have to be avoided or managed very carefully.

It is therefore inevitable that early spaceplanes will not be as safe as present airliners. A realistic initial safety target to aim for is that achieved by airliners in the 1930s when commercial flying first became a large-scale operation. In those days, there were approximately 10,000 flights per

▲ **Figure 11.1 de Havilland Comet jet airliner experimentally fitted with rocket engines to boost take-off performance.**

fatal accident, which compares with more than 1 million flights today. With continuous improvement resulting from in-service experience, the safety of spaceplanes will improve and eventually approach, if not quite reach, that of airliners.

Evidence of the potential safety of rocket-powered airliners is provided by an experiment carried out in 1951 and 1952. A de Havilland Comet jet airliner was fitted with rocket engines to boost take-off performance (see Figure 11.1). It made twenty flights, including three in one day. No particularly difficult safety problems were anticipated or found with the rocket engines. In the event, the jet engine performance was boosted sufficiently to avoid the need for the rockets.

In addition to the hazards of ascent and descent in spaceplanes, there are also risks due to being in space itself. Radiation from the Sun and more distant bodies is higher in space than on Earth because the atmosphere serves as a shield. Roughly speaking, a week in a spacecraft in normal conditions results in a similar dose of natural radiation as a year on Earth.

Potentially fatal doses occur during solar flares, which therefore present a major hazard, especially as they are irregular and unpredictable. However, they can be seen erupting from the Sun at least 15 minutes (usually much longer) before their radiation reaches Earth. The solution is therefore to use Sun-watching instruments that will give notice of forthcoming intense radiation, allowing crew and passengers time to be evacuated back to Earth or to retreat to highly shielded 'citadels' where they will be safe until the hazard has subsided.

Travelling further out from the protective magnetic fields of the Sun and Earth will expose crews to galactic cosmic radiation, and this could become a major problem for travel to Mars and beyond. Unlike radiation from solar flares, this is a long-term problem. While it may be possible to build radiation shields that are sufficiently light, we do not yet understand the full scope of the problem. However, this is an area of active research, and it is probably less a question of whether a solution can be found than one of how much it will cost in terms of added weight.

Space debris is a general term given to the many particles in space that are not recognized astronomical objects. It derives from two main sources: meteorites from the depths of the Solar System, and the remains of satellites and rockets that have collided or exploded. Such debris moves fast and can seriously damage a spacecraft.

Some debris is large enough to be detected, and this can in principle be destroyed or deflected by space-based weapons, or avoided by moving the spacecraft out of the way (the International Space Station has done this several times). Most debris is too small for this defence, and some shielding will probably be needed. Space hotels could be built in modules that, in an emergency, would be self-contained, each with equipment essential for survival. In the event of serious damage to one module, passengers and crew would evacuate to an undamaged one.

Long-duration space flights have shown that months of zero-g can cause medical problems such as bones becoming brittle. Tourists on a visit of a few days will not be affected. For longer stays, and for crew, and for voyages to more distant parts, a solution is to provide sufficient artificial gravity in a rotating section of the spacecraft to avoid these problems.

Another vexing problem is how to avoid contaminating Mars and other celestial bodies with terrestrial micro-organisms, and vice versa. Robotic probes are subjected to intensive decontamination, which would not be practicable for us humans.

In summary, spaceplanes will greatly improve the safety of space travel. Even so, it is likely to be significantly riskier than terrestrial travel for some time to come. Nevertheless, safety issues are unlikely to restrict progress greatly, at least as far out as the Moon. Beyond that, we have to find lightweight shields against cosmic radiation and solve the decontamination problem.

12

Roadmap to the new space age

So far in this book we have seen that the new space age will make a big difference to science and the environment, and that it will enable many of us to visit space. We have also seen that it can be achieved by developing projects that were considered feasible in the 1960s and that the major obstacle is mindset. This chapter considers in more detail how to bring about the new space age soon and at low cost and risk.

Spaceplane development strategy

Recapping from Chapter 4, a large spaceplane using mature developments of existing technology is capable of reducing the cost of sending people to space by around 1000 times, and is therefore well able to support the new space age. However, such a vehicle has a development cost probably beyond the reach of private sector investment. However, a 50-seat capacity is larger than needed for early spaceplane operations, which suggests that the *first* orbital spaceplane should be smaller. As discussed in Chapter 5, the first orbital spaceplane is the key to the new space age as it will enable the beneficial downward cost spiral to begin. Some of the 1960s spaceplane projects described in Chapter 3 were quite similar in overall concept to the large mature spaceplane described in Chapter 4 (see Figure 4.3), but were mostly significantly smaller, having a payload of around six astronauts. A 1960s-style spaceplane would therefore make a good lead-in to the larger vehicle.

Even a small orbital spaceplane is probably beyond the means of private investors. This suggests an even less ambitious project, within the means of the private sector, as the next step towards the new space age. A small suborbital spaceplane would meet this requirement admirably. This could be based on a 1950s rocket fighter.

It is therefore reasonably obvious that the development strategy (roadmap) for the new space age should start with a small suborbital spaceplane designed with features that are in line with later orbital spaceplane developments. This should be followed by a small orbital spaceplane, to be followed in turn by a larger one. This strategy is shown in Figure 12.1.

The next few sections of this chapter consider these vehicles in greater detail, starting with an assessment of the design requirements.

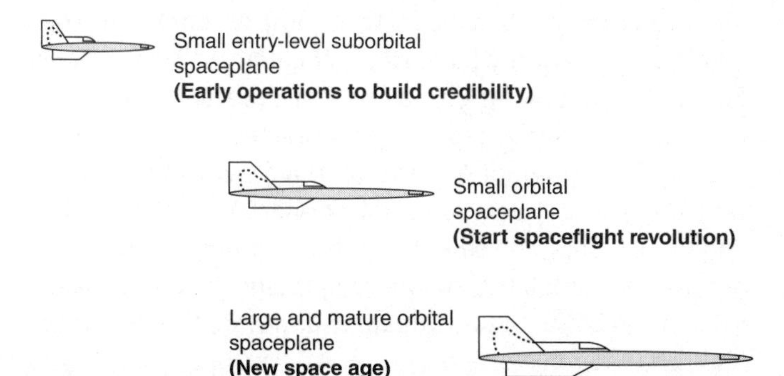

▲ **Figure 12.1 Spaceplane development roadmap.**

Design requirements

Of the twelve 'new space projects' mentioned in Chapter 3, nine involve launchers, and these are all very different. This is in line with a historical trend. Over the years, there have been numerous serious reusable launcher studies (probably more than 100), offering a wide variety of proposed design features. Some have had one stage; others more. Some have used wings for landing; others have been recovered by parachute, rotor, vertical jet or vertical rocket. Some projects have been piloted; others unpiloted. Some have taken off horizontally; others vertically. Some have used all-rocket power, while others have used jets for the early part of the ascent. In the 1960s, there was a near-consensus on some of the major features of one type of reusable launcher – the spaceplane – but there have been numerous proposals since then.

There is a useful comparison with the pioneering phase of aeroplane development. Within a few years of the Wright Flyer, just about every conceivable configuration for an aeroplane had been tried out, as shown in the 1965 British comedy film, *Those Magnificent Men in their Flying Machines*. After only a few more years, a particular configuration emerged as market leader. This was the fixed-wing biplane made of wood, wire and fabric, with a tail at the back and a propeller(s) in front of the engine(s). This so-called tractor biplane dominated aeronautics until the metal monoplane began to take over in the 1930s. Many 'dead-end' developments were built and then discarded.

Given the large number of spaceplane studies that have been carried out, it should be possible to avoid comparable dead-end developments by using straightforward design logic to derive a competitive set of top-level design features. A good starting point is to consider the requirements. First, development cost and risk should be as low as practicable. The major space agencies seem unlikely to undergo a Damascene conversion anytime soon, and the initiative will have to come from the private sector, with government support later. This requires a good business plan, which in turn means keeping cost and risk down. This requirement should not be difficult to meet. As we have seen, orbital spaceplanes were widely considered feasible in the 1960s, and all the required technologies have since been demonstrated in flight.

The next requirement is subtler. The invention of the spaceplane will lead to far greater numbers of people going to space. Safety will therefore have to be far better than can be achieved with expendable launchers. As already mentioned in Chapter 11, the Space Shuttle suffered 2 fatal accidents in just 135 flights. A large airport handles that number of flights in just a few hours, so that level of safety would be out of the question for aviation. The first spaceplane with the clear potential to approach airliner safety is likely to be the most attractive to the largest new market – carrying passengers – and so should become market leader. Most passengers offered a choice between a fully reusable piloted spaceplane that looks and flies like an airliner, or a spacecraft launched by an expendable vehicle, or one that lands using a parachute or vertically on a pad using rocket engines, or one that is not piloted, are going to choose the former. The top-level design goal should therefore be 'safety soon'.

Leading design features

We have already seen in Chapter 4 that spaceplanes using existing technology and designed for large new markets have some unavoidable design changes compared with airliners. They must have two stages, use some hydrogen fuel, use rocket engines, have reaction controls and thermal protection, and have a shape suitable for high speed. They must retain the airliner feature of full reusability. This still leaves a variety of optional design features to be settled.

The requirement for safety means that the design of the first orbital spaceplane should be as much like an airliner as is practicable, because airliners are the safest flying machines yet invented. The requirement for development soon means that advanced technology should be kept to a minimum. This section uses these requirements to derive some of the key design features of the first orbital spaceplane and the suborbital lead-in project.

Considerations of safety dictate that take-off and landing should be horizontal, using wings to provide lift. This is safer than vertical take-off or landing using rotors, jets, parachutes or rockets. Conventional aeroplanes have a better safety record than helicopters or vertical take-off jets. If engines are used to provide lift as well as thrust, it is more difficult to deal with the effects of engine failure.

Safety considerations also dictate that all stages should be piloted, as pilotless aeroplanes are far less safe and are likely to remain so for many years to come.

The above design features are all but essential for a spaceplane designed for safety in the near future. The remaining three features are less essential but are highly desirable.

First, the lower stage should be equipped with jet and rocket engines. Jets are more practical than rockets for taxiing, ferry flights, aborted landings, or for diverting to other airfields. Additional rockets enable the stage separation speed and height to be increased without needing advanced jet engines that are not yet available.

Second, the maximum speed of the jet engines on the lower stage should be supersonic, between, say, Mach 2 and 4. The faster the lower stage can go, the smaller is the propellant weight fraction needed by the upper stage. It can then use less advanced technology and/or carry more payload. The lower stage jet engines should therefore be as fast as practicable without requiring advanced technology, which is in the supersonic range.

These arguments explain why the large mature spaceplane described in Chapter 4 (see Figure 4.3) has all of these features.

Third, in order to keep costs down, the first orbital spaceplane should be no larger than needed to capture much of the early markets for spaceplanes. This will be mainly launching small satellites and supplying crew and supplies to space stations. This means a payload in the 1-tonne class, which is comparable with that of a medium van or pick-up truck. This is the only one of the above features not shared by the large mature spaceplane, which is designed to minimize cost when mature rather than pioneer new markets. In summary, the basic design features for the *first* orbital spaceplane should be as follows.

- Two stages, both piloted, and taking off and landing horizontally using wings to provide lift.
- Jet and rocket engines on the Carrier Aeroplane stage.
- Jet engines with a maximum speed in the supersonic range.
- A payload in the 1-tonne class.

Design example

Several of the 1960s projects came close to having all of the features described in the last section. What was then the 'least unfeasible' approach is now the most straightforward and is still the most competitive. Perhaps surprisingly, the only project on offer today that has the essential design features described above, let alone all of them, is the Bristol Spaceplanes *Spacecab*, shown in Figure 12.2.

The Carrier Aeroplane has roughly the same size and shape as Concorde but is much simpler. It has to keep up its maximum speed for just the minute or two needed for the Orbiter to separate, whereas Concorde had to do so for more than two hours. It uses existing jet engines up to Mach 2 and rockets for a burst up to Mach 4 and to a height where the air is thin enough for straightforward separation. The Orbiter then separates and carries on to orbit. The Carrier Aeroplane flies back to base ready for the next flight. It can launch a satellite in the 1-tonne class, or carry six astronauts or passengers to space stations.

This proposal for the first orbital spaceplane is in effect an update of the 1960s European Aerospace Transporter

▲ **Figure 12.2 The Bristol Spaceplanes *Spacecab*, designed to be the first orbital spaceplane.**

project, designed to keep development costs low by using existing technology.

Timescales

We saw in Chapter 3 that spaceplanes were widely considered feasible in the 1960s and that all the technologies for an orbital spaceplane have since been demonstrated in flight. Orbital spaceplanes are therefore now within the scope of a straightforward engineering development programme. Operational prototypes could therefore start carrying payloads about seven years after the start of design, which is a typical time between go-ahead and entry into service for an advanced aeroplane. Thereafter, it will be a question of maturing the design to reduce operating costs.

When it comes to achieving design maturity, the most difficult engineering task will probably be developing rocket engines that have useful lives of thousands of flights compared with the present tens. Chapter 3 described how it took eight years to mature jet engines from early operations to the first airliner service. Given comparable priority, there is no obvious reason why rocket engines should not approach maturity in a comparable timescale. Thus, mature spaceplanes could be flying within about 15 years – 7 years for operational prototypes plus 8 years for maturing the technologies. So, spaceplanes based on 1960s designs offer the prospect of a 1000 times launch cost reduction within 15 years!

Development cost

Development cost is a key issue. For a rough preliminary estimate, we can use the rule of thumb that the cost of developing new aeroplanes of comparable technology is approximately proportional to their empty weight. SpaceShipOne (SS1) cost about $30 million to develop, which is remarkably low for an aeroplane with a maximum speed of around Mach 3 that is capable of carrying three people to space. Scaling by weight, the development cost of the first orbital spaceplane comes out at just under $2 billion. This is clearly an approximate estimate. However, it is a small number compared with the roughly $8 billion per year that NASA and the European Space Agency between them spend on human spaceflight.

The first orbital spaceplane would save much of this cost by offering far less expensive transportation, which would also be far safer than space travel at present. So, if the large space agencies adopted spaceplanes, they would save money within a few years on present programmes alone, even including the cost of developing the spaceplanes themselves. The main challenge is persuading them to take this prospect seriously.

This $2 billion development cost might seem low, considering that a large new airliner costs several times that much. The difference is that a new airliner has to compete in a mature and competitive market in which a few percentage points of efficiency can make or break the project. By contrast, pioneering spaceplane developers will be creating a new market and setting standards that later vehicles will have to compete with. Early spaceplanes can therefore be built in relatively small experimental workshops (skunk works), which costs much less than developing fully certificated aeroplanes in the conventional manner. This is one reason why SS1 cost so little to build. Another is that it was built by an experienced and efficient team.

Suborbital lead-in project

As mentioned earlier, finding the funding for even a small orbital spaceplane is not going to be easy. Development costs are probably too high for private entrepreneurs, and space agencies are not showing much if any interest in spaceplanes. This suggests the need for an even smaller and less expensive lead-in project, within the scope of the private sector. A small suborbital spaceplane would meet this requirement. A rocket-powered aeroplane with suborbital performance and about the size of the X-15, SS1 or the SR.53, could be used for carrying passengers or science experiments on brief trips to space. It would thereby build up the market for space tourism and low-cost space science. It would also start to mature the technologies needed for an orbital spaceplane. The resulting gain in credibility for spaceplanes should enable the funding to be found for the first orbital spaceplane.

As far as practicable, the basic design features of this suborbital lead-in project should be the same as the first orbital one. Therefore, the essential features are that it should be fully reusable, should have wings, should take off and land horizontally, and it should be piloted. There is a choice to be made about the number of stages. Single-stage suborbital spaceplanes require a propellant weight fraction of around 60%, depending on engine efficiency, and are feasible with existing technology. However, air-launch from a carrier aeroplane improves performance at the expense of cost.Of the twelve private sector 'new space' projects under development, mentioned earlier in Chapter 3, just two are funded suborbital spaceplanes with these essential features – the Virgin Galactic SpaceShipTwo (SS2) and the XCOR Lynx.

Desirable but not essential features are that it should have the smallest useful size, which means a pilot and one passenger or scientific experiment. In the interests of practical and safe flying, and being in line with later developments, it should have jet engines in addition to the rockets. It should be clearly seen as the next step on the evolutionary path to the first fully orbital spaceplane, and so should avoid 'dead-end' technologies. Incredible as it may seem, the SR.53 rocket fighter, also mentioned in Chapter 3, comes close to meeting all of these requirements for a suborbital spaceplane, and that aeroplane first flew in 1957.

Other developments

Spaceplanes will be limited in size, and hence payload, by the need to use conventional runways. Large spacecraft can be assembled in orbit from modules carried in multiple spaceplane flights. For some very large payloads, it may be more economical to build a large reusable heavy-lift vehicle that would take off vertically to avoid the size limitation of using runways. This could be a development of an existing large launcher using spaceplane technology to provide reusability.

Roadmap overview

Summarizing this chapter so far, the first orbital spaceplane capable of approaching airliner safety while avoiding difficult new technology is likely to become market leader. The resulting design driver of 'safety soon' leads to the essential features of two fully reusable stages, both having wings and pilots and taking off and landing horizontally, plus several less essential but nonetheless desirable features. The first orbital spaceplane is the key project for the new space age and could be based on the best of the 1960s designs, several of which had these essential features. A small suborbital spaceplane is the natural lead-in project to the first orbital spaceplane, and this could be based on a rocket fighter from the 1950s. A larger and more mature orbital spaceplane can be built to follow on from the first one when the market is ready for such a vehicle. This in turn can be followed by a spaceplane using advanced jet engines to enable a single stage to be used.

How you can make a difference

From the discussion so far, we can conclude that there is a reasonably obvious way ahead for space transportation. The technology and markets are in place. Government space agencies could use existing budget streams to pay for spaceplane development, and would save money on existing programmes alone by so doing. The outstanding requirement is a change in the mindset that orbital spaceplanes have to be difficult and expensive.

As mentioned earlier in this chapter, two suborbital spaceplanes are under development – the Virgin Galactic SS2 and the XCOR Lynx. The success of either of these, or of as yet an unfunded competitor, would most likely lead eventually to the new space age. However, not one of the new space companies mentioned in Chapter 3 has presented a clear evolutionary path from its present programmes to an orbital spaceplane with features even approaching those assessed earlier in this chapter as being essential for success. The field is therefore wide open for someone to make a start along this roadmap, with a vehicle specifically designed to trigger that process.

In this chapter, I would like to invite you to indulge in a thought experiment in which you, with a very small team, trigger the revolution in spaceflight by breaking the mould of thinking. All you need is an open mind, the ability to delegate and, probably for most of you hardest of all, a budget of a few million dollars of your own or someone else's money.

Let us assume that you are not quite convinced by the arguments in the previous chapters that orbital spaceplanes could and should have been built in the 1960s, that they could now be built with existing technology, and that they could reduce costs by approaching 1000 times within 15 years. You think to yourself that if it were that simple, surely the big space agencies would be developing orbital spaceplanes with some enthusiasm, or one of the new US space companies would be building one, wouldn't they?

Yet you are open to persuasion. You are interested in space and appreciate its importance. You are puzzled by the lack of public discussion of radical space futures. You don't quite understand what has gone wrong, but you are of a practical disposition and would like to do something about it anyway. So you decide to make a start by learning more about aeroplanes.

You therefore decide to build a small aeroplane from a kit. You might decide to be adventurous and select the world's smallest jet – the FLS Microjet (see Figure 12.3), which is a sort of resurrected Bede BD-5J. You may remember seeing James Bond doing great things in this aeroplane in the film *Octopussy* (1983).

The FLS Microjet kit would set you back just under $200,000 and would take typically 2000 working hours to build. These numbers are high by homebuilt aeroplane standards because of Microjet's complexity. (Most homebuilt aircraft use small piston engines, have a fixed landing gear, and do not have flaps.) You find that you need do-it-yourself skills well above average and the ability to pay unusually careful attention to detail, so you team up with one or two experienced aeroplane kit builders and make a success of building and flying it. You then decide to impress your friends by designing your own little aeroplane. You decide to cut a bigger dash by using the same wing shape as the Space Shuttle Orbiter to make it look spaceworthy. So you add a designer or two

▲ **Figure 12.3 The FLS Microjet, available in kits for homebuilders.**

to the team, or learn how to design a light aeroplane, and the result might look like the drawing in Figure 12.4 (the Bristol Spaceplanes *Microsonic* is shown here because it is the only exemplar, at least in the public domain, suitable for present purposes).

You enjoy showing off in your little spaceship and ask yourself, more out of curiosity than expectation, what it would take to go supersonic. After all, your spaceship has the right general shape. You find that a smallish rocket engine and propellant tanks are all you need to add, so you end up with an aeroplane like the one in Figure 12.5. You cannot locate a suitable rocket engine in an aircraft industry catalogue, but you find that some engines built in the 1960s and 1970s would be ideal, and that re-engineering one of these would not be too difficult. You will also have to pressurize the cockpit and/or buy a pressure suit,

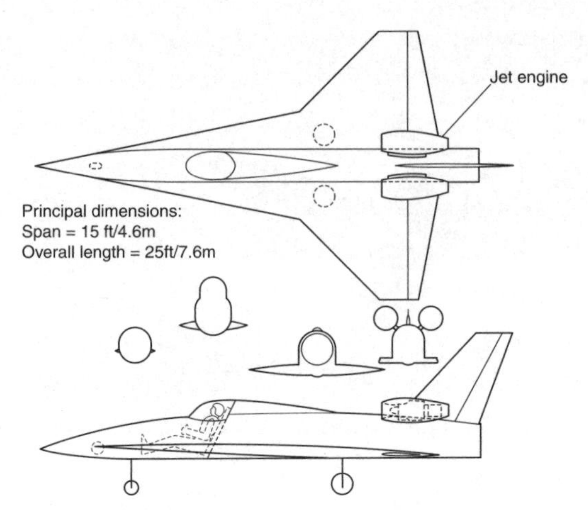

▲ **Figure 12.4 A small homebuilt jet aeroplane.**

and find a way of coping with rapid changes in control forces as the aeroplane passes through the speed of sound, but you find that these are not insuperable obstacles. You will however need to employ some professional aero engineers, because the engineering is now quite advanced.

Your design calculations tell you that if you tried to fly your little aeroplane at supersonic speed at low level, the air loads would tear the wings off (unless the structure were so strong as to be impracticably heavy). You avoid this by flying high enough for the air to have a sufficiently low density for the loads on the aeroplane to be reasonably low. (Air loads decrease directly as the air density decreases with height, and are roughly proportional to the speed squared). You find that the required height for supersonic speed with air loads acceptable to a typical

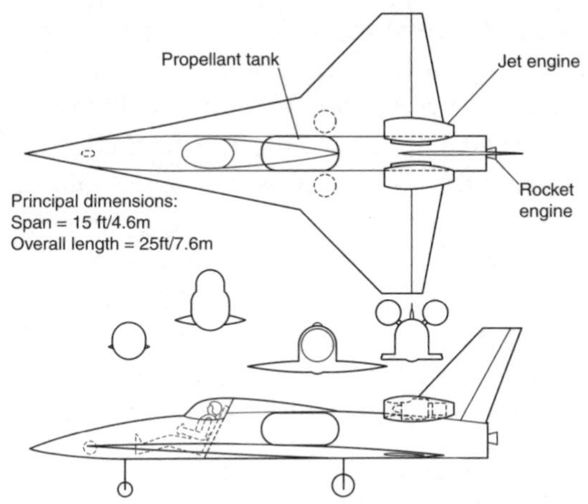

▲ **Figure 12.5 A small homebuilt supersonic jet and rocket aeroplane.**

light aeroplane structure is about 55,000 ft (17 km). You can reach this height more readily with a rocket than with a jet. As the height increases, the thrust of a rocket engine increases slightly but the thrust from a jet varies roughly with air density and at 55,000 ft (17 km) is only about one-eighth of that at sea level.

Having got into the *Guinness Book of Records* for building the world's smallest supersonic aeroplane, you look for the next record. You find that by fitting a larger propellant tank and a larger rocket you can gain the world height record for aeroplanes taking off under their own power, which at present is the 123,523 ft (38 km) gained in 1977 by Alexandr Fedotov in a MiG 25M.

Having broken two records, you get carried away and ask yourself, ever so tentatively, what you would have to do to reach space height. To your amazement you find that, if you remove the jet engines to save weight, fill the fuselage with propellant, fit a larger rocket, and get towed by an aeroplane to 9 km (30,000 ft), you can indeed zoom-climb up to 100 km (328,000 ft), which counts as space. True, you will have to add reaction controls for use outside the atmosphere and a thin layer of thermal insulation around the nose and the leading edges of the wing, but again these design challenges are not insuperable. You will, however, have to build up a team of a dozen or so aircraft professionals. Your aeroplane might now look like the one in Figure 12.6.

You are now in the *Guinness Book of Records* yet again for having built the world's smallest spaceplane. This will have cost you several million dollars. You will have shown that spaceplanes do not have to be difficult or expensive.

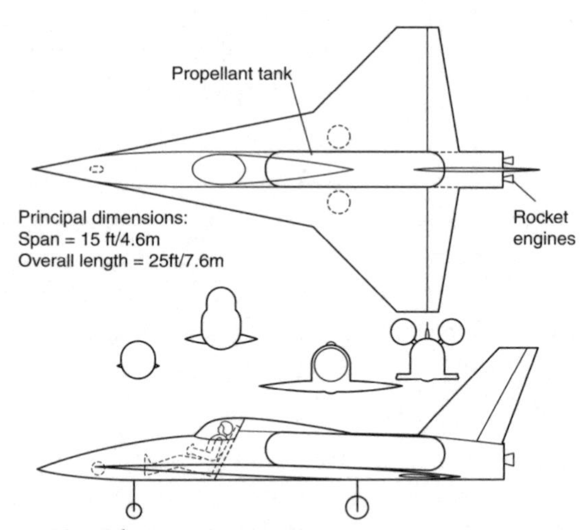

▲ **Figure 12.6 A 'homebuilt' spaceplane.**

Some may argue that your little aeroplane is only suborbital and that the real problems will hit you when you try to go orbital. However, you are able to reply (you are now quite an expert) that you can build something like the first orbital spaceplane described earlier with the lower stage like a much enlarged version of your small spaceplane and the upper stage also somewhat enlarged and fitted with an (existing) hydrogen-fuelled rocket engine and a thicker layer of surface insulation to deal with the heat of re-entry, as first demonstrated by the Space Shuttle in 1981. This, of course, will require the resources of a medium or large aircraft company.

You use your new-found fame to draw attention to the roadmap discussed in this chapter which, you are now convinced, is the way to go. Having demonstrated that the idea that spaceplanes have to be exceptionally difficult and expensive is a myth, there is a very good chance that serious backers and industrial partners will emerge and that you will indeed have brought forward the revolution in spaceflight.

It is possible for an individual to make a big difference in this way because the basics of low-cost access to space were worked out in the 1960s and then not acted upon. 'All' that you have to do is to draw attention to this in a way that is likely to lead to further action.

13

Finally

The dominant barrier in the way of expanded space exploration is high cost. This book has presented a roadmap for slashing costs soon and at low technical risk. The main requirement is to replace expendable launchers like missiles with reusable ones like aeroplanes. It is likely that the cost of sending people to space can thereby be reduced by about 1000 times in around 15 years.

The roadmap involves catching up with the opportunity that was missed in the 1960s to develop spaceplanes. Projects that would then have been feasible but expensive to develop would now be straightforward. The biggest obstacle at present is no more than, and no less than, mindset. Fifty years of using expendable launchers has led to the widespread belief that spaceplanes require advanced technology and a very large budget for development. The 1960s design consensus has been largely overlooked or forgotten. Progress has been stalled since the 1980s because the Space Shuttle as built was not fully reusable.

Three insights have been emphasized. First, when it comes to routine transportation for large markets, a reusable launcher is as profound an advance over an expendable one as an aeroplane is over a balloon. Expendable launchers cannot fly more than once and balloons cannot fly into wind, and so neither type of vehicle can break out of niche markets. By overcoming these fundamental limitations, spaceplanes and aeroplanes can so break out, and both are suitable for large markets. The invention of the aeroplane triggered an explosive growth in air transport. Likewise, the invention of the orbital spaceplane will trigger an explosive growth in space transport.

Second, if the best ideas from the 1960s are adopted, the first orbital spaceplane, which is the pivotal development for the new space age, can be built soon and at modest cost and risk.

Third, space tourism is likely to be the first market to become large enough to generate the funding, economies of scale, and enthusiasm needed to develop spaceplanes to their full potential soon.

Spaceplane development will lead to three fundamental changes. First, the change from throwaway to reusable launchers and to an aviation approach to space transportation. Second, the change in government responsibility from being in the lead to supporting and regulating commercial organizations, as is the case with today's air travel. Third, the change in spaceflight markets to dominance by public access for business and leisure, again as with today's air travel. These changes add up to nothing less than a re-conceptualization of our approach to space activities.

The benefits of low-cost access to space will be widespread. In the near future, there can be a new golden age of astronomy and space exploration. Public access to space stations for business and leisure can soon become widely affordable. There can be wide environmental benefits. If space tourists react like astronauts to date, there could even be a boost to creative thinking about the future of our planet.

The advantages of spaceplanes are so great that developing one soon is in effect **all that matters** at present for space exploration.

Not everyone is going to agree with these conclusions, but if this book contributes to an open public debate it will have served its purpose.

100 IDEAS

Films

1 *2001: A Space Odyssey* (1968)

2 *Apollo 13* (1995)

3 *Those Magnificent Men and their Flying Machines* (1965)

4 *Destination Moon* (1950)

TV series

5 *Wonders of the Universe* BBC Two, 19 February 2012

6 *Wonders of the Solar System* BBC Two. Series started on 4 April 2010

Books

7 *The Right Stuff* by Tom Wolfe (Farrar, Straus and Giroux, 1979)

8 *The Emperor's New Clothes* by Hans Christian Andersen (First published in *Fairy Tales Told for Children*. First Collection. Third Booklet. 1837)

9 *Our Final Century – Will the Human Race Survive the Twenty-First Century?* by Sir Martin Rees, then Astronomer Royal and now President of the Royal Society (William Heinemann, 2003)

10 *The Singularity is Near* by Ray Kurtzweil (Viking, 2005)

11 *Countdown* by T. A. Heppenheimer (Wiley, 1999)

12 *V-2* by Walter Dornberger (Viking Press, New York, 1954)

13 *The Cambridge Encyclopedia of Space* (Cambridge University Press, 1990)

14 *Wonders of the Solar System* by Professor Brian Cox and Andrew Cohen (Harper Collins, 2010) This is the book of the TV series mentioned above.

15 *The High Frontier: Human Colonies in Space* by Gerard K. O'Neill, (New York: William Morrow & Company 1977)

16 *2081: A Hopeful View of the Human Future* by Gerard K. O'Neill (New York: Simon and Schuster 1983)

Websites

17 www.spacefuture.com (Webmaster: Peter Wainwright)

18 The Lurio Report by Charles Lurio. www.thelurioreport.com

Things to expect

19 The first peace conference in orbit – will the new perspective encourage compromise?

20 Space debris detection and avoidance systems

New jobs

21 Space traffic controllers

22 Solar flare forecasters

23 Space research scientists and technicians

24 Spaceplane pilots and mechanics

25 Spaceplane cabin attendants

26 Space hotel staff

27 Space hotel designers, architects and interior designers

28 Space fashion designers

29 Spaceplane maintenance technicians

30 Safety regulators

31 Propellant suppliers

32 Space insurers

33 Spaceplane designers and production workers

34 Space doctors and nurses

Orbital hotels

35 **Phase 1** – Single module, non-rotating

36 **Phase 2** – multi-module, non-rotating

37 **Phase 3** – rotating structures providing artificial gravity

38 **Phase 4** – very large structures for sports stadia, zero-g gyms, swimming pools

Space sports

39 Zero-g judo

40 Zero-g wrestling

41 Zero-g flying

42 Zero-g swimming

43 Zero-g table tennis

44 Orbital Olympics

Culture and tourism

45 Orbital art studios

46 Lunar tourism (technically straightforward once orbital tourism starts, but may turn out to be less attractive than tourism in Earth orbit). Visits to historic sites, e.g. Apollo, Lunokhod remains)

47 Space tourism centres

48 Space travel clubs

49 Educational outreach

Other facilities

50 Orbital apartment blocks

51 Orbital hospitals

52 Orbital chapels

53 Orbital 'graveyards'

54 Re-entry cremations

Movers and shakers

Note: This is a list of people whose work has most influenced the author. It is certainly not intended to be a complete or authoritative record of those who have contributed most to the new space age.

55 **Burt Rutan** – aeronautical engineer par excellence and founder of the Scaled Composites company, which built the first privately funded spaceplane, SpaceShipOne (SS1)

56 **Buzz Aldrin** – the second man to step on the Moon, and long-time advocate of reusable launchers

57 **Charles Lurio** – perhaps the most effective journalistic advocate of the new space age and recorder of progress towards its fulfilment (see 'Websites' section above)

58 **Dennis Tito** – the first space tourist. By paying $20 million and undergoing six months of tough astronaut training, Tito demonstrated that the market for space tourism is an exceptional one

59 **Elon Musk** – founder of SpaceX, which is re-writing the rules for launcher development by building the Falcon series at a fraction of the traditional cost

60 **Eric Anderson** – co-founder of Space Adventures, which has arranged for seven private citizens to visit the International Space Station. He is also co-founder of Planetary Resources, which was set up to explore the feasibility of mining asteroids

61 **Eugen Sänger** – German rocket pioneer. His studies of the 1930s and 1940s led to the European Aerospace Transporter projects of the 1960s. In 1933, Sänger published a report, *Raketenflugtecknik* (*Rocket Flight Engineering*), that was translated into English and Russian and became very influential. It contains a concept sketch of a rocket-powered research aeroplane that looks remarkably like the X-15 which flew 26 years later

62 **Neil Armstrong** – the first man to step on the moon

63 **Patrick Collins** – a long-standing advocate of space tourism. Pioneer of space tourism market research and the economics of solar power satellites

64 **Paul Allen** – co-founder of Microsoft, funder of SS1 and founder of Stratolaunch

65 **Peter Diamandis** – founder of the International Space University, the X-Prize, and the Singularity University.

The X-Prize led directly to SS1, the success of which has made the new space age all but inevitable in the next decade or two.

66 **Ray Kurtzweil** – inventor and creator of the 'Singularity' concept to describe what may happen when we invent computers that exceed our mental capacity in all important aspects

67 **Richard Branson** – the first entrepreneur to set up an airline, Virgin Galactic, offering space experience flights

68 **Robert Bigelow** – the first entrepreneur to build space stations for commercial use

69 **Thomas F. Rogers** – one of the first to appreciate the importance of space tourism as a means of lowering the cost of access to space. His foundation provided the first seed money for the X-Prize foundation

70 **Wernher von Braun** – the prime mover of the V-2, the first vehicle to space, and the Saturn rockets that took the first men to the Moon

71 **Yuri Gagarin** – the first man in space

Studies to do

72 **Astronomy:** What new astronomical instruments will become feasible with low-cost access to space? How will they help research?

73 **History:** Why were spaceplanes not developed in the 1960s, when they were first considered feasible?

74 **History:** What effect did the US victory in the race to the Moon have on Soviet opinion? Was the young Gorbachev, for example, impressed and did this influence his support for perestroika?

75 Social sciences: What will be the impact of space tourism on human global perspectives?

76 Policy: Spaceplanes could greatly reduce the cost of major government projects, so why are governments so slow to react?

77 Business: How can innovation theory be applied to spaceplane development?

78 Space tourism: How big is the likely market?

79 Environment: What will be the impact of spaceplanes?

80 Asteroid mining: What are the commercial prospects?

81 Solar power satellites: What should be the first pilot scheme after spaceplanes enter service and make such a project affordable?

82 Technology: What are the timelines for exploring the Solar System using nuclear rockets and electromagnetic catapults?

83 Innovation theory/science fiction: Considering unintended consequences, how soon before they first occurred were computer viruses, hacking, spam, the internet economy, and cyber warfare predicted? What might be the unintended consequences of low-cost access to space?

Quotations

84 Richard van der Riet Woolley FRS, Astronomer Royal, 1956, on space travel: 'It's utter bilge. I don't think anybody will ever put up enough money to do such a thing ... What good would it do us? If we spent the same amount of money on preparing first-class astronomical equipment we would learn much more about the universe ... It is all rather rot.' Woolley's protestations came just one year prior to the launch of Sputnik, five years before the launch of the Apollo programme, and thirteen years before the first landing on the Moon.

85 **Henry Ford, *Theosophist Magazine*, February 1930:** 'The time will come when man will know even what is going on in the other planets and perhaps be able to visit them.'

86 **Mahatma Ghandi:** 'First they ignore you, then they laugh at you, then they fight you, then you win.'

87 **Mike Mullane,** NASA astronaut, when asked by a NASA psychiatrist what epitaph he'd like to have on his gravestone, answered, 'A loving husband and devoted father', though he jokes in *Riding Rockets*, I would have sold my wife and children into slavery for a ride into space'. *Riding Rockets: The Outrageous Tales of a Space Shuttle Astronaut* R. Mike Mullane. (Simon & Schuster, 2006)

88 **Carl Sagan,** 'All civilizations become either spacefaring or extinct.' *From Pale Blue Dot: A Vision of the Human Future in Space* (New York: Random House, 1994)

89 **Stuart Atkinson,** 'Some say that we should stop exploring space, that the cost in human lives is too great. But Columbia's crew would not have wanted that. We are a curious species, always wanting to know what is over the next hill, around the next corner, on the next island. And we have been that way for thousands of years.' New Mars, 7 March 2003

90 **Konstantin Tsiolkovsky:** 'The Earth is the cradle of humanity, but mankind cannot stay in the cradle forever.'

91 **Buzz Aldrin,** 'I believe that space travel will one day become as common as airline travel is today. I'm convinced, however, that the true future of space travel does not lie with government agencies – NASA is still obsessed with the idea that the primary purpose of the space program is science – but real progress will come from private companies competing to provide the ultimate adventure ride, and NASA will receive the trickle-down benefits.' *Magnificent Desolation: The Long Journey Home from the Moon* Ken Abraham (Crown Publishing Group, 2009)

92 **Carl Sagan, *Pale Blue Dot*:** 'Imagine we could accelerate continuously at 1 g – what we're comfortable with on good old terra firma – to the midpoint of our voyage, and decelerate continuously at 1 g until we arrive at our destination. It would take a day to get to Mars, a week and a half to Pluto, a year to the Oort Cloud, and a few years to the nearest stars.'

93 **C. S. Lewis,** 'A man who has been in another world does not come back unchanged. One can't put the difference into words. When the man is a friend it may become painful: the old footing is not easy to recover.' *Perelanda* (The Bodley Head, 1943. Also titled *Voyage to Venus* in a later edition by Pan Books)

94 **David Grinspoon, *Slate*, 7 January 2004:** 'As long as we are a single-planet species, we are vulnerable to extinction by a planetwide catastrophe, natural or self-induced. Once we become a multiplanet species, our chances to live long and prosper will take a huge leap skyward.'

95 **Stephen Hawking, *The Daily Telegraph*, 16 October 2001:** 'I don't think the human race will survive the next thousand years, unless we spread into space. There are too many accidents that can befall life on a single planet. But I'm an optimist. We will reach out to the stars.'

96 **Ray Bradbury,** 'Space travel is life-enhancing, and anything that's life-enhancing is worth doing. It makes you want to live forever.' Playboy, 1996

97 **Buzz Aldrin, *Esquire*, January 2003:** 'In my mind, public space travel will precede efforts toward exploration – be it returning to the moon, going to Mars, visiting asteroids, or whatever seems appropriate. We've got millions and millions of people who want to go into space, who are willing to pay. When you figure in the payload potential of customers, everything changes.'

98 **Mary McCarthy, *The Paris Review*, winter-spring 1962:** 'It really seems to me sometimes that the only hope is space. That is to say, perhaps the most energetic – in a bad sense – elements will move on to a new world in space. The problems of mass society will be transported into space, leaving behind this world as a kind of Europe, which then eventually tourists will visit. The Old World. I'm only half joking.'

99 **Freeman Dyson, *Discover Magazine*, June 2008:** '[Space travel] will come, but only when there is a high enough demand so that you can have a "public highway" system. To support today's air traffic network, you've got to have a million passengers constantly on the move. The same will be true in space: It's not really a technology problem, it's more a sort of chicken-and-egg economic problem. I hope that it will grow, probably on the back of the military. The military has needs for all kinds of space launching and is prepared to pay for it. So with luck something like this space highway will develop. It doesn't matter who pays for it initially. In the end it will belong to everybody.'

100 **Thomas R. Lounsbury (1838–1915):** We must view with profound respect the infinite capacity of the human mind to resist the inroad of useful knowledge.

Glossary

Ballistic missile Used in the modern sense to refer to large rocket-powered guided weapons with no wings.

Carrier Aeroplane The aeroplane-like lower stage of a multi-stage launch vehicle.

Chemical rocket engine An engine in which fuel and oxidizer, both carried by the vehicle, combine chemically in a combustion chamber to generate a high temperature gas, which exits through a nozzle to produce thrust. 'Rocket' is often used loosely to describe a complete rocket-powered vehicle that takes off vertically and is expendable, like a rocket firework, ballistic missile or conventional launch vehicle.

Fuel The chemical substance that reacts in an engine with the oxidizer. In a rocket-powered aeroplane, 'fuel' is sometimes used loosely to mean fuel and oxidizer, i.e. all the propellant.

g Used here to mean a person's apparent weight divided by that at the Earth's surface. A person at the Earth's surface supported by a chair or floor, etc. will 'feel' 1 g. This 'feel' is due to the restraint preventing him or her from falling towards the centre of the Earth, pulled by gravity, at an acceleration of approximately 9.8 meters per second per second. In orbit in a non-rotating spacecraft, there are no restraining forces, and the person will feel weightless, or experience 'zero-g'.

Hypersonic Speeds faster than about Mach 5, or five times the speed of sound. At this and higher speeds, the shock waves are close to the aircraft surface, and aerodynamic characteristics do not change greatly between Mach 5 and orbital velocity, which is approximately Mach 25.

ISS The International Space Station

Mach number Aircraft speed divided by the local speed of sound. The speed of sound depends on local temperature, and is about 760 mph (340 m/s) at sea level and 660 mph (295 m/s) in the stratosphere.

Man-tended Used to describe a device that works most of the time robotically but which needs occasional maintenance or modification

Maturity A new engineering project typically progresses through a succession of phases, which might include design, prototypes, testing, pre-production, production, operational service, and product improvement, until design maturity is reached. The product should then be well understood, safe, easy to maintain, and reliable. A mature spaceplane should approach airliner standards of life, maintenance cost, and turnaround time.

NASA The US National Aeronautics and Space Administration. This is the world's largest space agency.

New space age This expression is increasingly used to denote space activities when present private sector initiatives reach fulfilment, especially greatly reduced transportation cost.

Orbit A trajectory of a small body in space around a large one, which is periodically repeated, such as a satellite around the Earth. Unless stated otherwise, it refers to low Earth orbit, i.e. a more or less circular orbit around the Earth at a height of a few hundred kilometres.

Orbiter The upper stage of a multi-stage reusable launch vehicle that reaches orbit.

Oxidizer The chemical substance consumed in an engine that reacts with the fuel. In a jet engine, the oxidizer is the oxygen in the air. In a rocket engine, the oxidizer is carried on board the vehicle in a separate tank. It is often liquid oxygen or a chemical containing oxygen.

Payload The weight of the useful carrying capacity of a vehicle, such as passengers, mail, weapons, research instruments, or cargo.

Propellant Fuel and oxidizer carried in a vehicle for use in a rocket engine. For convenience, when discussing jet aeroplanes, 'propellant' is used to describe just the fuel. The oxidizer is the oxygen in the atmosphere.

Reaction controls Small rocket motors used to change a vehicle's orientation when in space. (Ordinary flying controls do not work in the absence of air.)

Re-entry heating The heat caused by the friction of air when a vehicle re-enters the atmosphere from space.

Satellite speed This is the speed at which an object stays in a circular trajectory around a much larger one, like a satellite around the Earth. At this speed, the centrifugal force due to the circular motion balances the force due to gravity at the height in question. In low Earth orbit, satellite speed is about 5 miles/sec (8 km/sec).

Solid rocket booster A rocket booster is a lower stage that uses a solid rocket motor. A solid rocket motor uses a propellant in which the oxidizer and fuel are both solids. These are mixed together and squeezed into the rocket tube. For safety reasons, they are designed to require a high pressure and temperature to ignite, these being provided by an igniter. Display fireworks use solid rockets. They are heavier and less efficient than high-performance liquid-fuel rockets.

Space height As height above the Earth increases, the atmosphere becomes less dense. Space height is the height at which the atmosphere is deemed to stop and space to start. This is largely a matter of definition but, for aeronautical engineers, it is the height at which the air loads on a vehicle become very low. There are various definitions of space height. In the 1960s, the United States Air Force awarded Astronaut Wings to X-15 pilots who exceeded a height of 50 miles (80 km). At this height, the sky is dark with bright stars even in daytime. At this height, air loads start to become significant on the Space Shuttle as it re-enters the atmosphere from orbit. The Federation Aéronautique Internationale (FAI) defines the edge of space as 100 km (62 miles). The FAI is the international body regulating aircraft speed and other records, and its jurisdiction stops at this height.

Spaceplane Used here to mean an aeroplane-like fully reusable vehicle that can fly to and from space height and that lands using wings for lift. Unless prefixed by 'suborbital', this refers to a vehicle that can fly to and from orbit.

SS1 SpaceShipOne, built by the Scaled Composites company. In 2004 it became the first privately funded aeroplane to reach space height.

SS2 SpaceShipTwo. An enlarged development of SS1, being built for Virgin Galactic.

Stage Launchers are an assembly of vehicle elements, all but one of which are discarded during the acceleration to orbital velocity.

These elements are called stages. Each stage increases the velocity until orbit is achieved. A single-stage vehicle is theoretically possible, but has not yet been built.

Space Tug A space vehicle used to transfer objects from one orbit to another, such as from a low Earth orbit to a high one, or from an Earth orbit to a Lunar one.

Suborbital A trajectory that reaches space height but not at sufficient speed to stay up like a satellite. Unless prefixed by 'long range', this refers to a vehicle that spends only a few minutes in space and lands back at or near to where it took off.

Zero-g A convenient term to describe the weightless conditions experienced in orbit in a non-rotating spacecraft. Strictly speaking, the 'g' is nearly always more than zero, and 'microgravity' is a more precise term.

Zoom-climb A flight manoeuvre in which speed is traded for height. The aeroplane starts at high speed in level flight, the pilot pulls it up into a steep climb, and the speed drops off as height is gained. Zoom-climbs are a means of briefly reaching great height.

Index

1960s spaceplanes 23–7

accidents 104–7
aeroplanes, invention 56–7
Aerospace Transporter project 23–4
air travel 48–50
airliners 48–54
Apollo project 3–4, 21, 71, 90–1
Ariane 5 15, *16*, 43
asteroid mining 4–5, 69–71
astronaut fatalities 104
astronomy 11–13
atmospheric opacity *11*
atmospheric pollution 88

ballistic missiles 19–20
Bell X-1 22
Bigelow Aerospace 36
Boeing 37
Boeing 747 *49*
Brightman, Sarah 72

catapults, electromagnetic 95–6
chemical rockets 41, 68, 93, 107
colonization 93–102
contamination 110
cost
  high 47
  per flight 53–4
  reduction 48–50

Dassault Aerospace Transporter 25–6
design challenges, launch vehicles 41–7
design changes, to airliners 50–4
development strategy 112–13
Dream Chaser 36

EADS Rocketplane 37
'Earthrise' photograph 90–1
economic expansion 89
EELT (European Extremely Large Telescope) 64
electromagnetic catapults 95–6
employment 89
environmental science 88–91
Envisat satellite 88–9
European Extremely Large Telescope (EELT) 64
exoplanets 12
expendability 15, 47–8, 106
extraterrestrial intelligence 71

factories 65–6
fatalities 104
flight testing 105–6
FLS Microjet 126–7

'homebuilt' spaceplane 126–31
hotels 80–2

International Space Station (ISS) 13, 63

James Webb space telescope 63–4
jet development 29–30
jet engines 41
jobs 89
Juno 1 20

Kennedy, John F. 21
Kurtzweil, Ray 102

launch vehicles
design challenges 41–7
history 19–33
launchers, satellite 14–16
legal issues 84–6
long-range transport 73–4
lunar base 66–8
Lynx suborbital spaceplane 36, 37, *38*, 39

manufacturing 65–6
Mars 68–9, 93–4
medical problems, zero-g 110
Messerschmitt Me 163 rocket fighter 30
Microsonic 127, *128*
mining asteroids 4–5, 69–71
Moon race 21–2
Moon Treaty 84–5

NASA 28–9, 33–4
NF-104A rocket plane 31
nuclear rockets 94–5

Orbital Sciences 37
orbital spaceplanes 23–7, 57–9

Planetary Resources 4–5, 36, 69
Pohl, Henry O. 90–1
pollution 88
'posthumans' 100
pressure suits 77–8
probes 12–13, 62
propellant weight 41–3

radiation 108–9
Reaction Engines 37
Redstone 20
revolutions, in transportation 56–9
robotic probes 12–13, 62
rocket development 29–30
rocket engines 50–1
rocket-powered aeroplanes 22–3, 31–2
The Rocketpropelled Commercial Airliner 26–7
Russian Federal Space Agency (Roscosmos) 13

safety 104–10, 115–16
Sänger, Eugen 23–4
satellites 12, 14, 20, 26, 61–3, 88–9
Saunders Roe SR.53 rocket fighter 31–2
Scaled Composites 33, 35
Short-Mayo Composite 45, *46*
Sierra Nevada Corporation 36
single-stage launchers 43
Skylon reusable launcher 37
SLS (Space Launch System) 34
solar power 65, 98
sounding rockets 17
space colonization 93–102
space debris 109–10
space elevators 96–8
space exploration 68–9
space holidays 76–82
space hotels 80–2
Space Launch System (SLS) 34
space law 84–5
space probes 12–13, 62
space science 63–4
Space Shuttle 15–16, 25, 27–9, 33, 43, 105, 106
space sickness 78
space stations 13, 63
space tourism 26, 27, 71–3, 76–82
space towers 96–8
Spacebus 52
Spacecab 118–19

spaceplanes
 1960s 6, 23–7
 cost 53–4
 design 50–3, 114–16, 116–20
 development cost 121–2
 development strategy 112–13
 'homebuilt' 126–31
 revolution 57–9
 safety 107–10
 suborbital lead-in project 122–3
Spaceship Company 35
SpaceShipOne (SS1) *22*, 23, 32
SpaceShipTwo (SS2) 5, 36, 37, *38*
SpaceX 5, 35
Sputnik 20
SS1 (SpaceShipOne) *22*, 23, 33
SS2 (SpaceShipTwo) 5, 36, 37, *38*
steam locomotives 56–7
Stephenson's Rocket *56*, 57
Stratolaunch Systems 36
suborbital flights 16–17, 121–2
suborbital spaceplanes 44
Sud-Ouest Trident rocket fighter 31
swimming 81–2

technology development, rate of 101–2
telescopes 12, 63–4
'The Singularity' 102
Tito, Dennis 72
tourism *see* space tourism
tourist suits 77–8
training, for space tourists 77–8
transhumanist movement 100
treaties 84–5
two-stage launchers 43–6

United States Air Force Space Command 14

V-2 ballistic missile 19–20, 44–5
vehicle certification 85–6
Virgin Galactic 4–5, 36, 37, *38*, 39

Wright Flyer *56*

X-15 22–3
XCOR 36, 37, *38*, 39

zero-g
 medical problems 110
 tourist activities 81–2